AF605768

The Long Shadow

The Long Shadow

How Introduced Species Changed the Fauna of Aotearoa

Photography by
CALLY WHITHAM

To Greg, my wonderful, supportive husband, and Ethan and Willow, my patient children.

Bloodthirsty animals are now swarming over every part of the country, and I defy all attempts to check their increase. The intention, of course, was to find a new remedy for the wild rabbit nuisance, but it is notorious that these marauders will not take fur when they can get feather.

'The Vanishing Forms of Bird-Life in New Zealand', *The Press*, Volume LIV, Issue 9622, January 11, 1897.

Contents

Preface

'Most of us are observing introduced species when we point out birds, animals or trees to our children.'

Photography, history, and a love of birds all took root for me in childhood. I grew up surrounded by portraits of my ancestors, their stories passed down through generations, and a taxidermised kiwi inherited when my great-grandfather moved to a rest home. Each morning, I would lift the lid of its glass case to give it a gentle pat; a small ritual that tied me to the past.

My first photographs were of the animals on my great-aunt's farm. I thought farm animals were my main love, until one summer I watched a blackbird meticulously build a nest and raise her chicks. That was the moment I realised my fascination with birds.

Later, while studying photography, I found myself drawn to portraiture, influenced by the ancestral photographs that had been a constant presence in my life. Over the years, that influence has shaped my work, even as my subjects have shifted and expanded.

I returned to photographing introduced species when my children were small. I remember holding my son as a baby, pointing to a blackbird and saying 'bird' to him, realising this was the first one he'd ever seen. As he grew, 'bird' was naturally followed by 'cow', 'sheep', 'dog', 'cat', 'tree', etc. — the familiar species we encountered in our rural environment. I cannot pinpoint the moment I introduced him to a native bird or tree; those were rarer sightings on our daily walks or drives through farmland. This led me to wonder how many of us, when teaching our children to name the world, are introducing them mostly to species brought here by human hands.

With two young children in tow, my photography was often confined to whatever was close at hand: passerines, poultry, livestock and waterfowl. As my catalogue of images grew, so did my curiosity. Where did these species come from? How and why were they brought here?

The answers unveiled a story of determination and folly. Introduced species, initially brought for food, sport and sentimentality, became agents of ecological change on a scale no one had foreseen. What started as attempts to recreate a piece of 'home' quickly spiralled into a cascade of poorly judged introductions, with devastating consequences for Aotearoa New Zealand's unique environment.

This book emerged from my growing collection of portraits and the unsettling history I uncovered through old texts and newspaper accounts. The story of how our family came to possess a taxidermised kiwi — once a cherished curiosity — has since revealed its own uncomfortable truths emblematic of a broader, troubling legacy.

This is not an exhaustive catalogue of the introduced and native species of Aotearoa, but a visual narrative journey. Part 1 explores the introduced species, their origins and their transition into pests. Part 2 shifts to our native species, highlighting what has been lost and what remains under threat. Together, these images and stories capture the environmental toll of misguided sentimentality, shortsighted introductions and greed.

Through this work, I hope to prompt reflection: on what we've lost, what we must fiercely protect and what we might still regain — though never in full.

Cally Whitham

Foreword

'This book so clearly describes the "old lady who swallowed a fly" scenario, where they unleashed introductions to deal with previous introductions that resulted in irreversible unintended consequences for our native wildlife.'

'Ka mua, ka muri' is an oft used whakataukī that loosely means to walk backwards into the future. The idea of this proverb is that to understand where we're going, we've got to have a proper sense of where we have come from. This book has been published at a fascinating time in our history — we New Zealanders are wrestling with our identity as we continue to grow as a nation together. One of the constant characters which defines our national identity is our wildlife — to which we are inextricably linked. We pay homage to our indigenous species in myriad ways, like pouring funding into our native flora and fauna: from kiwi to kōkako, kārearea to whio. RNZ National once tried to remove the daily bird call that breaks up the hours of the morning news reports — to abject horror and rejection by the local public. Yet, as Cally rightly points out in the preface of this book, many of us don't 'know' indigenous wildlife as part of our daily lives; instead, we are familiar with the foreign. It begs the question: if intruders now define us, what does that mean for who we are? To lose any of our indigenous species (and we have lost many already) is, in my view, to lose instruments from an orchestra, eventually tamping down the harmonies to lonely, discordant notes.

This book is an oddity in the best way. Its stunningly beautiful photos (I struggle to remember they *are* photos, as they look like they have been lovingly rendered with brushstrokes on canvas) are a signpost to our past, whereby Cally has staged her subjects as though they are in a Goldie painting. Even today, many of us continue to gaze upon portraits of our ancestors in living rooms, museums and marae. Cally has directed our gaze to the wildlife that surrounds us by placing them in portraiture in such a unique way. To complement her breathtaking photography, she has also dug through Aotearoa archives and captioned her photos with actual proceedings of various journals and media from the time. She has done so to tell the story of what we already know: that while the onslaught of animals brought by acclimatisation societies were familiar to settlers from home, they didn't belong here. For example, at a time when there appeared to be just one dunnock remaining, there is a report in 1922 of 'many people journeying to the gardens to see the stranger.'

These remarkable photos of our endemic wildlife are, in my view, a warning: that if we don't get to know our indigenous animals as well as we do the exotic, they might also be consigned to mere portraits in a museum. It behoves us all to reflect on our unique endemic wildlife in these beautiful images and to read history's portrayal of what they meant to the new cultures who reached the shores of Aotearoa.

While early European settlers often dismissed our native ecosystems with derision, this book so clearly describes the 'old lady who swallowed a fly' scenario, where they unleashed introductions to deal with previous introductions that resulted in irreversible unintended consequences for our native wildlife, the culmination of which must surely be the stoat. We must learn from these early failed experiments, ensuring that our precious taonga — those very creatures which define us all — are not lost to future history books. To lose our species forever is to lose the very essence of our identity.

We are Kiwi, after all.

Nicola Toki, chief executive, Forest & Bird

Introduction

> The progress of acclimatisation, already so successful, will gradually replenish this portion of the earth with every domestic animal and bird profitable and useful to man. It will stock our woods with game and our rivers with fish, while bringing the feathered songsters of our mother-country to delight the ears of our children with their sprightly melody.

Inaugural address of Governor Sir G. F. Bowen, *Transactions and Proceedings of the Royal Society of New Zealand* 1, (1868):1.

It is doubtful whether the people of Canterbury today realise what they owe to the men of a past generation in regard to the beauties of plantation and garden; the fascinations of splendidly-stocked trout stream and river; the plentiful supply of sport with gun and dogs; and the charms of abundant bird life in field and shrubbery. Few indeed pause to think how these welcome accessories to the everyday pleasures of life came to be here in such lavish profusion. They are accepted as part of the attractions of the locality and are taken very much as a matter of course. The enterprise that brought them here, the worry, the disappointments, the anxieties, the expense, are not thought of.

It was not until 1864 that some of the early settlers began to seriously feel the want of a number of things that had formed part of their life at home, and, as a result, a movement was initiated to form an acclimatisation society.

A circular was issued pronouncing that 'It is proposed to establish a society, having for its object the introduction of beasts, birds, fishes, and vegetable productions of such species as will be acclimatised with probable advantage to this district and the colony.'

'The Story of our Plants and Birds,' *The Press*, Volume LVII, Issue 10840, December 15, 1900.

'Men went forth with slaughtering weapons in their hands, and the overflowing scourge of devouring fire was sent throughout the country till a great part of the flora was consumed, and the birds had to seek food in remote places.'

It was with the advent of Europeans that destruction began in earnest. It seemed as if they had been commanded to destroy the ancient inhabitants. Men went forth with slaughtering weapons in their hands, and the overflowing scourge of devouring fire was sent throughout the country till a great part of the flora was consumed, and the birds had to seek food in remote places.

The first European animals were introduced by Captain Cook. On his second voyage, in 1773, he let loose three pigs in Queen Charlotte Sound. They afforded food and sport for Māori, and also for the adventurous Europeans who lived in the colony in its early civilised days. They also took part in the war of extermination, and it was probably owing to their depredations that the tuatara was banished from the mainland.

An inkling of the full effects that were to follow the advent of civilisation was given when the whalers visited these waters and established stations on the shores. Their special mission was the destruction of the whales, but they made their presence felt in other ways, notably by the introduction of dogs, cats and rats, before which many of the flightless birds fell easy victims.

Then the settlers arrived with fresh supplies of domestic animals and deadly firearms. By shooting, the numbers of the avocets, white herons in the south, and the pigeons, ducks, swamp hens, quails and other birds were greatly reduced. In some favoured spots, the slaughter was terrible.

Burning grass, scrub and bush wrought further havoc, and very soon the quails were almost destroyed, while no crows were left on the eastern side of the South Island. Dogs killed the wood hens in large numbers. Lands were cleared of timber; swamps were drained, and the native flora was largely supplanted by crops, hedgerows and gardens, and by Old World shrubs and trees. The landbirds retreated in diminished numbers, mostly to the mountains, where the ancient forests, still standing, offered them a place of refuge. New animals were introduced systematically. These greedy invaders soon overcame any feeble resistance that may have been offered and speedily established themselves.

Some of the new animals were brought in by Europeans for food, and some for purely sentimental reasons. There were few natural food supplies in New Zealand when colonists first came.

'The cultivation of cabbages, cauliflowers, turnips and other succulent plants was followed by an alarming increase in the numbers of native insects. Armies of caterpillars invaded the fields and consumed the crops.'

It became obvious that if the colony was to be the abode of a large population of civilised beings, acclimatisation must take place on a large scale. Food supplies were, therefore, the first consideration. Sheep, cattle and other animals found a climate resembling that of their native country and pastures which could not be surpassed. Like their predecessors, the 'Captain Cooks', they throve very well and grew fat. Large cattle and sheep runs were formed, and when the needs of the settlers were satisfied, there was established an export trade in wool and mutton that has attained to great dimensions.

Rabbits, pheasants, the honeybee, and later on, quails, hares, deer, and trout and other food fishes were also introduced.

The cultivation of cabbages, cauliflowers, turnips and other succulent plants was followed by an alarming increase in the number of native insects. Armies of caterpillars invaded the fields and consumed the crops. It was hardly possible to open a pea-pod without finding a caterpillar inside, and, in the Auckland district, dismayed settlers saw fields of maize under bare poles, not a leaf remaining. The food supply of the insects had increased enormously, and they were not slow to respond.

It was decided that the best plan to adopt, to make agriculture and horticulture possible, was to introduce insect-eating birds. But it was recognised that these birds must not live on insects alone. There is no winter retreat for insect-eaters in New Zealand as there is in Europe, and if they could not sustain themselves on vegetable food in the winter months when the insects were absent, they would perish. The field of selection was therefore restricted to birds that would eat both seeds and insects, which would not try to migrate, and which would become common.

On the whole, the results of acclimatisation in New Zealand must be considered favourable, although unfortunate mistakes have been made. Among these, the most marked is the introduction of rabbits, about which little need be said, as the disastrous effect of their presence is too well known, and of ferrets, stoats and weasels.

F. W. Hutton and J. Drummond, *The Animals of New Zealand: An Account of the Dominion's Air-breathing Vertebrates*, (Whitcombe and Tombs, 1909).

Part 1

Exotic

Aves

One hundred and forty-four bird species were introduced by man between 1840 and the present day; of these, thirty-three have survived and are part of New Zealand's avifauna.

The following are the only immigrants who have become truly wild: mallard, Canadian goose, black swan, common pheasant, Chinese pheasant, Australian swamp quail, Californian quail, rock pigeon, little owl, skylark, thrush, blackbird, dunnock, Australian magpie, rook, starling, myna, house sparrow, chaffinch, redpole, goldfinch, greenfinch, cirl bunting and yellow-hammer; a total of twenty-four species. The record of failures is much greater than that of successes.

G. M. Thomson, *The Naturalisation of Animals & Plants in New Zealand*, (Cambridge University Press, 1922).

Insectivores

An army of caterpillars, hundreds of thousands strong, was overtaken by a train as the insects were crossing the rails to reach a field of oats. Thousands of them were crushed under the wheels of the engine, and the train suddenly stopped. It was found that the wheels had become so greasy that they revolved without advancing as they could not grip the rails.

J. Drummond, 'On Introduced Birds,' *Transactions and Proceedings of the Royal Society of New Zealand* 39, (1906).

With the wholesale destruction of the bush to make farmland, ecology was so disrupted that the country was overwhelmed by plagues of insects which 'crawled over the lands in vast hordes'. The gathering of caterpillars was a sight that caused consternation to agriculturists. 'They came not in regimes and battalions but in mighty armies, devouring crops as they passed along and leaving fields totally bare as if there had never been seeds sown.'

A Dunsandel farmer says: 'I once saw caterpillars coming out of one man's paddock and crossing the road into another paddock. I made all haste to tell the man, and we got about 1600 sheep on the road and killed the insects.' In the same province, a settler who was driving his dray along a road drove through a colony of caterpillars which happened to be crossing the road at the same time. They were present in such countless numbers that the wheels of his dray were in a puddle, caused by the crushing of hundreds of insects.

The introduction of many insectivorous birds, including quails, pheasants, starlings, mynas, species of *Fringillidae* and others, must also have accounted for the destruction of immense numbers of insects. At one time, the wide tussock-covered hills and plains of both the North and South Islands were alive with grasshoppers, and in summer months they sprang up before the pedestrian literally in the thousands.

J. Drummond, 'On Introduced Birds,' *Transactions and Proceedings of the Royal Society of New Zealand* 39, (1906).

ABOVE RIGHT

REDPOLL

Acanthis flammea

The abundance of redpolls are shown from the following: 'In March 1911, Mr T. H. Jones of Christchurch caught seventy in one pull of the net near New Brighton; Mr C. Bills of Dunedin said he could complete an order for 1000 of them within a fortnight; while in Southland, as many as 500 were taken in a day.'

G. M. Thomson, *The Naturalisation of Animals & Plants in New Zealand*, (Cambridge University Press, 1922).

ABOVE LEFT

DUNNOCK (HEDGE SPARROW)

Prunella modularis

The Canterbury Society liberated nine in 1868, and forty-one in 1871. Mr Drummond says (1907): 'It was Captain Stevens who brought the first hedge sparrow to the colony, and it is claimed to the Southern Hemisphere. It came in the *Matoaka* together with the first house sparrows. It was the only survivor of a consignment. For a long time, it was an object of interest in the Society's grounds in Christchurch, with many people journeying to the gardens to see the stranger.'

G. M. Thomson, *The Naturalisation of Animals & Plants in New Zealand*, (Cambridge University Press, 1922).

OPPOSITE

HOUSE SPARROW

Passer domesticus

The late Mr Bills used to narrate how he trapped the sparrow for the Canterbury Society with large folding nets in the streets of London in the early mornings, and how Londoners were surprised that any country should want such birds. He explained that the caterpillars in New Zealand were so numerous and large that the farmers had to dig trenches around their houses to trap and bury the voracious creatures, lest after eating all the crops, they should turn and eat the farmers themselves.

J. Drummond, 'On Introduced Birds,' *Transactions and Proceedings of the Royal Society of New Zealand* 39, (1906).

OPPOSITE

CHAFFINCH
Fringilla coelebs

The difficulty of properly harvesting the seed, owing to the depredations of those unmitigated pests the chaffinch and greenfinch, proteges of the Acclimatisation Society — that rash and, unfortunately, irresponsible body with which this part of the colony has been afflicted and who had such a splendid opportunity of tilling our groves and hedgerows with beautiful, comparatively harmless and very beneficial feathered friends, but have instead placed curses upon this colony for all time.

C. T. Wren, *New Zealand Herald*, Volume XIX, Issue 6563, November 29, 1882.

ABOVE

EUROPEAN GOLDFINCH
Carduelis carduelis

There were no native finches, goldfinches or any other kind in New Zealand before they were brought here by man. They arrived in 1862. First introduced to Nelson, they spread thence over various paths across the country. It is doubtful whether the introduction thus made was altogether desirable; for casting a living creature upon new conditions always means a good deal of risk. So far, the goldfinch has taken the best path, being both beautiful and fairly harmless as it goes about its life in the countryside. They can blend a number of notes into a very sweet song.

Alex C. Begg, 'Councillor Robin and the Acclimatisation Society's Ducks', *Otago Daily Times*, Issue 7317, July 29, 1885.

ABOVE

SONG THRUSH
Turdus philomelos

The Otago Society introduced two thrushes in 1865, four in 1867, forty-nine in 1868, forty-eight in 1869 and forty-two in 1871. There was no mistake as to the determination of the Otago settlers to have their favourite songbird settle in New Zealand. It shows, too, the hardiness of this bird in confinement at sea.

J. Drummond, 'On Introduced Birds,' *Transactions and Proceedings of the Royal Society of New Zealand* 39, (1906).

OPPOSITE

EURASIAN BLACKBIRD
Turdus merula

The blackbird soon became naturalised. Colonists only smiled when it took a little fruit. There was plenty of fruit, they said, but there were only a few blackbirds, and they looked upon the bird's depredations as they would look upon the small failings of a favourite child. As the years went by and the blackbird increased in numbers, it began to take the lion's share of cherries, strawberries, pears, apples and other fruit. Gardeners then began to look upon it as an ugly, sooty intruder and a greedy nuisance, and its company was found to be not half as desirable as had been anticipated.

J. Drummond, 'On Introduced Birds,' *Transactions and Proceedings of the Royal Society of New Zealand* 39, (1906).

OPPOSITE

COMMON MYNA

Acridotheres tristis

This bird, like the sparrow, follows man in his migrations. Upon the approach of autumn, mynas largely use the roads, closing in on homesteads for scraps of fowl feed and the leavings of the gallows and the kennels.

This species appears to have been introduced in the first instance in all centres by private individuals and by a few of the societies in the early 1870s. One of the most remarkable things about them is their increase after their first introduction, and then their subsequent diminution, and — winsome districts — their ultimate disappearance. The latter appears to have been due, either directly or indirectly, to the starlings; the increase of the latter coinciding with the decrease of the former.

G. M. Thomson, *The Naturalisation of Animals & Plants in New Zealand*, (Cambridge University Press, 1922).

ABOVE

COMMON STARLING

Sturnus vulgaris

William Morrin, Esq., who returned to Auckland yesterday in the SS *Lord Ashley* after a visit to England, brought with him a very valuable selection of birds in the interest of acclimatisation. The following have been received — four golden pheasants, two mandarin ducks, two Carolina ducks, sixteen partridges, twenty-two starlings, twenty-four blackbirds and thrushes, thirty linnets and sixteen skylarks. They have arrived in capital order, having experienced remarkably fine weather throughout the voyage from England. This route is admirably adapted for bringing birds out from the old country.

'Acclimatisation', *Daily Southern Cross,* Volume XXIV, Issue 3343, April 3, 1868.

EASTERN ROSELLA
Platycercus eximius

We learn from the *Hobarton Mercury* on the 4th March that Mr Graves, the honourable secretary of our Acclimatisation Society, shipped by the *Wild Wave*, for the kindred society at Dunedin, a pair of the brush kangaroo and about 200 parrots, of the rosella kind, under the special and gratuitous care of Captain Fisher.

Hawke's Bay Weekly Times, Volume 2, Issue 70, May 4, 1868.

ABOVE
AUSTRALIAN MAGPIE
Gymnorhina tibicen

They are readily tamed and very sociable in confinement, but they are apt to drop their beautiful song and take to imitate all sorts of domestic sounds. I knew of one in Dunedin that could bark like a dog and mew like a cat, but its favourite amusement was to sit on the fence and call the fowls together.

G. M. Thomson, *The Naturalisation of Animals & Plants in New Zealand*, (Cambridge University Press, 1922)

OPPOSITE
LAUGHING KOOKABURRA
Dacelo novaeguineae

Mr Seddon introduced a laughing jackass from Australia, and as everyone knew, that bird lived on young snakes when in the bush. It would have been thought that a bird which could tackle snakes would be good enough for a stoat or weasel, but it had been demonstrated that such was not the case, as his valuable bird had been destroyed at his residence in Molesworth Street by a weasel.

'How the Premier Lost his Jackass', *The Press*, Volume LVIII, Issue 11020, July 18, 1901.

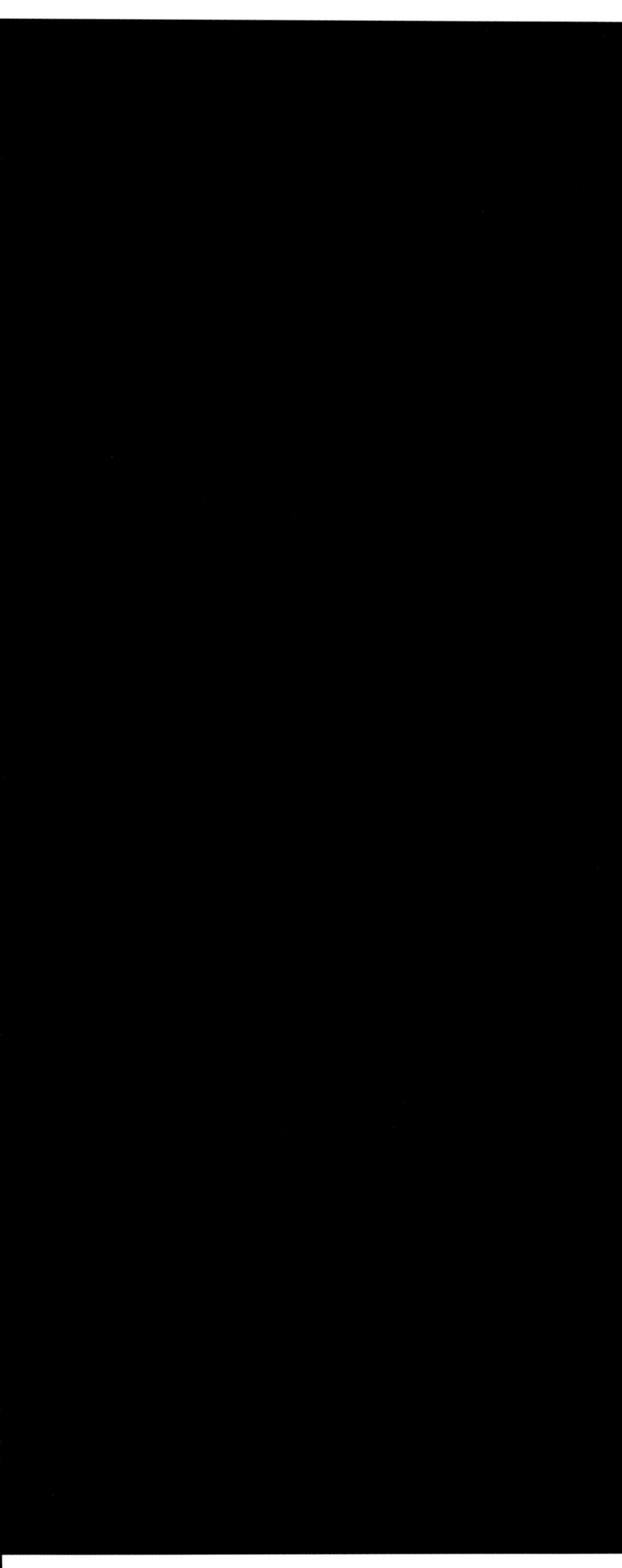

ROOK
Corvus frugilegus

A rookery would be a most cheerful addition to the New Zealand country house. The bird is a 'black coat' who for one tythe he takes of the farmer's crop, takes fifty tythes of the 'insect pests' that injure the farmer's crop; and there are worse things in the world than cold squab pie.

Charles Hursthouse, *New Zealand, or, Zealandia, the Britain of the South, Vol. 1,* (Edward Stanford, 1857).

LITTLE OWLS
Athene noctua

The little owl made its appearance near Invercargill in the autumn of 1915. I frequently heard its call but did not see the bird till 21st September, when I discovered a pair perched in a hole in a dead kāmahi. The little owl is much more difficult to approach than the morepork; it evidently sees much better in daylight, but apart from this, it seems to be a more wary bird. Unlike the morepork, the little owl hoots vigorously in the middle of the day — at least in the spring. It is now pretty common in this district; a few evenings ago, I heard four or five calling in different directions.

G. M. Thomson, *The Naturalisation of Animals & Plants in New Zealand*, (Cambridge University Press, 1922).

Ornamental & Game

The average (so-called) sportsman is a man out to kill something, and he does not concern himself with the amount of trouble and expense which has been gone to in order to provide him with the thing to be killed.

G. M. Thomson, *The Naturalisation of Animals & Plants in New Zealand*, (Cambridge University Press, 1922).

TOP RIGHT
CHUKAR
Alectoris chukar

ABOVE
ROCK PIGEON
Columba livia

Perhaps the most singular postal service in the world is the Carrier-Pigeon Post of New Zealand, between Auckland and Great Barrier Island. The training of the pigeons for the journey from Auckland to the Island, which lies low and is often hidden in fog, seemed at first a hopeless one. The pigeons refused to leave the mainland for the long sea passage of sixty miles across the Hauraki Gulf, but by patience and perseverance, this difficulty was at last overcome.

The pigeons upon their arrival push two wires apart with their shoulders and enter a little compartment two square feet in size. The wires in recoiling strike an electric bell, which summons an attendant, who takes the message from the birds and puts them into the dovecote.

'Carrier-Pigeon Post: A System Known to the Ancients', *Mataura Ensign*, Issue 871, March 28, 1901.

BOTTOM RIGHT
GREY PARTRIDGE
Perdix perdix

We regret exceedingly to find that the attempt of the Acclimatisation Society to introduce hares and partridges from England in the *John Bunyan* has proved a sad failure. Out of forty hares and forty-eight partridges put on board, only one hare and four birds have reached us alive. The vessel encountered bad weather in the early part of the voyage, and the boxes in which the game were placed were not waterproof, to which we must attribute this very serious mortality. One of the birds is a red-legged partridge, the only one of the kind shipped, the remaining three being the common grey partridge. We have not learnt the sex of the latter.

'Provincial and General', *North Otago Times*, Volume XIII, Issue 490, February 4, 1870.

ABOVE

BROWN QUAIL

Synoicus ypsilophorus australis

To the Editor of the *New Zealand Mail*. Sir, — I see by an advertisement in your paper that Mr W. T. L. Travers offers, on behalf of the Acclimatisation Society, a reward of £5 for information leading to the conviction of any person shooting quails, etc. Now, I have several times tried to kill some quails, but being an impatient sportsman and an indifferent shot, I cannot manage it. Quails are positively as great a nuisance as the rabbits, and unless the Acclimatisation Society compensates settlers for the loss of grass seed and other damage, they have no right to take them under their protection. Besides, a farmer can shoot any wild fowl on his own property without permission of Mr Travers or any other body — Yours, &c., Cockatoo.

'Protection of Game,' *New Zealand Mail*, Issue 481, April 30, 1881.

OPPOSITE

CALIFORNIA QUAIL

Callipepla californica

I wrote to Mr John Pollock, President of the Nelson (Acclimatisation) Society, on the subject, and in his reply, dated 13th December, 1915, he says: 'With regard to your enquiry re: quail I have interviewed Mr Kirkpatrick and he informs me that he started tinning these birds about twenty-five years ago, and from my own recollection that is about the time they were most numerous in this district. Mr Kirkpatrick informs me that the year he started the industry, he was paying as low as threepence per brace for the birds which were, of course, trapped, the price rising each successive year until ten pence was reached. At this price, it did not pay him to continue, and he accordingly closed down this line of goods.'

G. M. Thomson, *The Naturalisation of Animals & Plants in New Zealand*, (Cambridge University Press, 1922).

OPPOSITE

COMMON PHEASANT
Phasianus colchicus

To the Editor of the *Daily Southern Cross*. Sir, — Mr Ranger Ryland has a great deal to say in favour of the pheasant. Will he dare to affirm that he is ignorant of the immense amount of mischief committed among the growing crops this season by the pheasant? I am sure in his travels, he must have constantly heard the wailings of the settler regarding the depredations done by this bird. If by some extraordinary accident he has not, I shall be glad to inform him, through your paper, of instances almost innumerable of the utter ruin of grain crops brought about by this pest. We used to look upon the caterpillar visitation as the greatest curse that could befall us, but I have heard many say they would much sooner have that than that the pheasant should ever have been introduced. The former came once in three or four years; the latter is at work all the year round — morn, noon and night. More than one of the immigrants in my district has said to me that the pheasant is worse than a pig in a garden.

'Pheasant: Its Destructive Habitats,' *Daily Southern Cross*, Volume XXIV, Issue 3275, January 15 , 1868.

ABOVE

HELMETED GUINEA FOWL
Numida meleagris

The Rev. R. Taylor states that guinea fowls were first introduced by the early missionaries, who brought them to the Bay of Islands.

In the North Island, private individuals liberated them at several points, but they do not seem to have established themselves commonly as wild birds. I am informed, however, that in the Aberfeldy district, about forty miles inland from Whanganui, they are not uncommon. Mr Holman — curator of the Whangārei Acclimatisation Society, tells me that guinea fowls attack and drive away harrier hawks.

G. M. Thomson, *The Naturalisation of Animals & Plants in New Zealand*, (Cambridge University Press, 1922).

PEACOCK
Pavo cristatus

According to E. Jerningham Wakefield, the first peacocks introduced into New Zealand were brought to Wellington in 1843 by Mr Petre, who imported a large assortment of stock and materials in a ship of his own chartering. The Otago Society introduced two in 1867 and handed them over to some private individual, but no further record was kept of them. In other districts, private individuals and dealers introduced them, and, especially in the North Island, they occasionally got into the bush and became wild. Mr Thos. Hallett tells me that they were formerly wild in several places in Hawke's Bay, but disappeared. They are still wild in bush districts inland from Whanganui. Mr W. W. Fulton states similarly that they were formerly numerous in the valleys of the Turakina and Whangaehu Rivers.

G. M. Thomson, *The Naturalisation of Animals & Plants in New Zealand*, (Cambridge University Press, 1922).

MALLARD
Anas platyrhynchos

In spite of the fact that the native wild duck of New Zealand is as fine a bird, both for sport and table purposes, as any species of the family that can be introduced, the various acclimatisation societies have for many years made continuous efforts to naturalise other species, and notably the mallard.

The Otago Society got a pair in 1867 from the Melbourne Society and later from London introduced five in 1869, four in 1870, three in 1876, and nine in 1881. Apparently, none of these early introductions throve. It took many years of diligent and careful breeding to get the breed to establish significantly enough to allow shooting.

G. M. Thomson, *The Naturalisation of Animals & Plants in New Zealand*, (Cambridge University Press, 1922).

CANADA GOOSE
Branta canadensis

In 1905 the Government imported a considerable number of these birds and distributed them widely, eleven going to the Southland Society, ten to Otago, a number (unspecified) to Canterbury and six to Wellington. Mr Ayson reports in 1915 that they are doing well in several parts of the dominion. I am told this species is a migrant in its original habitat, in which case its establishment in New Zealand is rather interesting.

G. M. Thomson, *The Naturalisation of Animals & Plants in New Zealand*, (Cambridge University Press, 1922).

OPPOSITE

BLACK SWAN
Cygnus atratus

The black swan settled down at once to its new conditions. It was introduced into Canterbury partly with the object of destroying the watercress in the Avon, which runs through Christchurch. In a few years, the birds had increased largely, but in 1867 many of them forsook the Avon and made long and rather notable migrations to the wild country on the West Coast, and to Otago and even Marlborough.

J. Drummond, 'On Introduced Birds,' *Transactions and Proceedings of the Royal Society of New Zealand* 39, (1906).

ABOVE

MUTE SWAN
Cygnus olor

There is a very striking contrast between the white swan and the black swan with respect to their acclimatisation in New Zealand. The black swan is near the top of the list of success, while the white swan has increased slowly and with obvious difficulty and has sometimes quite failed to establish itself.

J. Drummond, 'On Introduced Birds,' *Transactions and Proceedings of the Royal Society of New Zealand* 39, (1906).

BLACK SWANS
Cygnus atratus

Because of the results and efforts of the Acclimatisation Society, I may state that black swans are now breeding in the lagoon by the brewery. The other day, a poor unfortunate goose wandered too near the haunt of the swan, and suffered penalties of intrusion on its domestic duty of hatching its offspring by getting its eyes pecked out and tasting other punishment.

Oamaru Bruce Herald, November 21, 1873.

Domestic

There were in the early days many wild fowls in the bush in Pigeon Bay, but whether or not they were descended from Captain Cook's stock I cannot say. It is much more likely that they were introduced either by whalers or by the early settlers at Akaroa and Lyttelton.

G. M. Thomson, *The Naturalisation of Animals & Plants in New Zealand*, (Cambridge University Press, 1922).

Captain James Cook's second voyage from 1772–1775 marked the initial introduction of domestic birds to New Zealand. Cook's primary objectives were scientific discovery, mapping new territories and studying natural history. In 1773, he brought domestic fowls, intending to provide his crew with fresh meat and eggs on extended journeys. Although his journals give few specifics on the types of birds, these early introductions set a precedent for future arrivals of foreign species, initiating ecological changes that would shape Aotearoa New Zealand's landscapes for years to come.

OPPOSITE
BARNYARD SPECIAL
Gallus gallus domesticus

ABOVE
BARRED PLYMOUTH ROCK
Gallus gallus domesticus

TOP LEFT
WYANDOTTE

TOP RIGHT
POLISH

BOTTOM LEFT
WHITE LEGHORN

BOTTOM RIGHT
SILKIE

In West Bay, Captain Cook liberated some fowls; in his later visit in February 1777, he stated that 'all the Māori whom I conversed with agreed that poultry are now to be met with wild in the wood behind Ship Cove; and I was afterwards informed by the two youths who went with us that Tuitou, a popular chief among them, had a great many cocks and hens in his separate possession.'

G. M. Thomson, *The Naturalisation of Animals & Plants in New Zealand*, (Cambridge University Press, 1922).

TOP LEFT
ORPINGTON BLUE

TOP RIGHT
GOLDEN CAMPINE

BOTTOM LEFT
RED LEGHORN

BOTTOM RIGHT
ORPINGTON

The common hen makes the best mother for the most varieties of fowls. Her want of discernment or indifference to the pedigree of her supposed progeny make her a valuable ally in raising those odds and ends of poultry which most people have about their barns at one or other period. Ducks, geese, turkeys, guinea, peacocks, etc. are all alike; the subject of her affectionate solicitude, and probably she would own a toad were one placed under her at hatching time.

Patea Mail, Volume VIII, Issue 957, November 10, 1882.

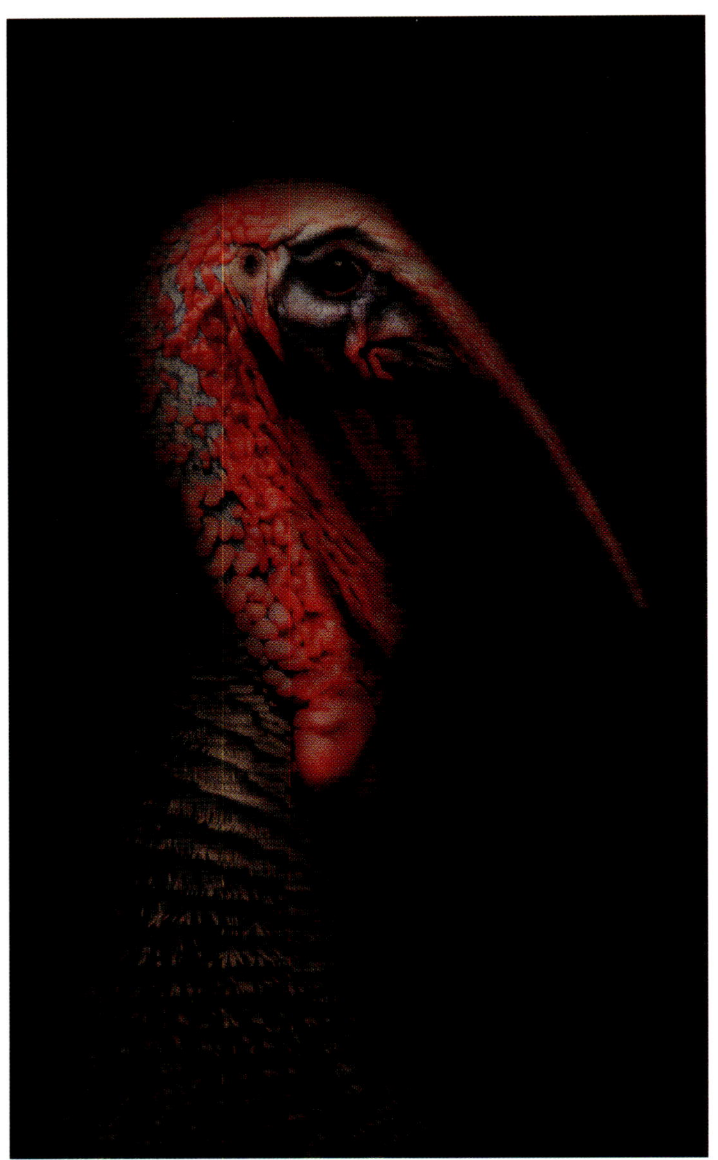

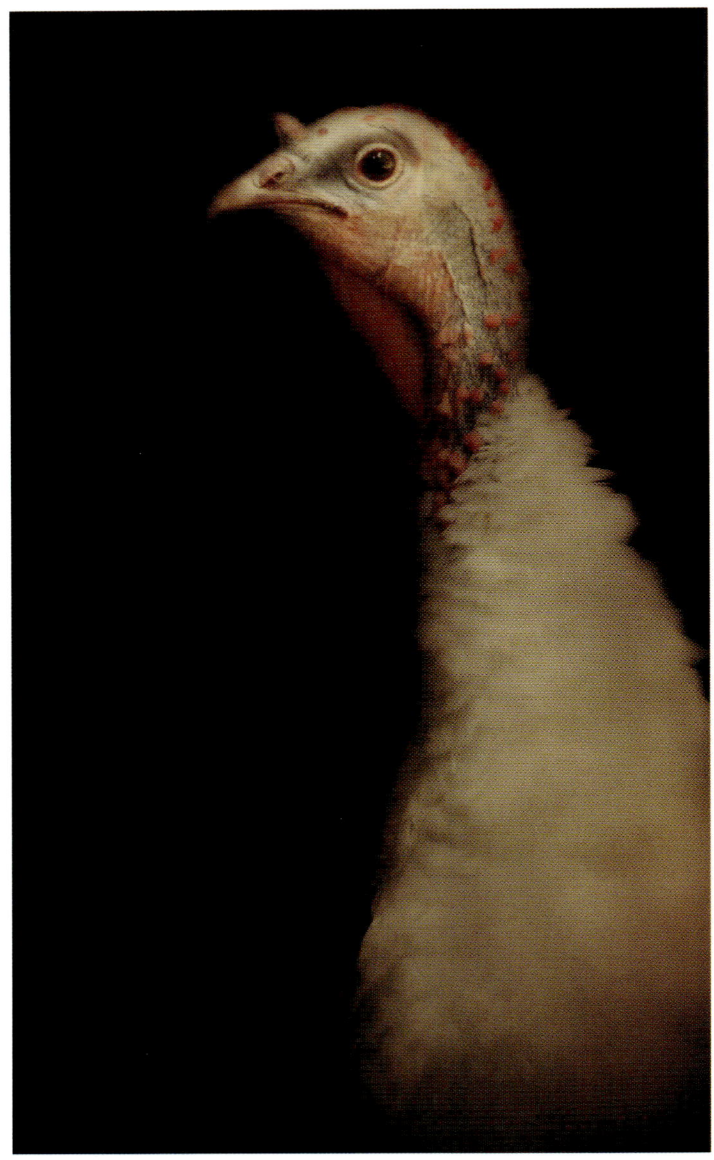

ABOVE LEFT

WILD TURKEY

Meleagris gallopavo

ABOVE RIGHT

WHITE TURKEY

Meleagris gallopavo

OPPOSITE

ROYAL PALM TURKEY

Meleagris gallopavo

A plague of caterpillars is reported in our local paper as troubling the settlers in the Bay of Plenty. For this, a simple and profitable remedy is available: keeping mobs of turkeys and moving them to wherever the pest develops and spreads . . .

Large areas of rape and turnips were grown, and the caterpillars made terrible havoc with the crops until the insectivorous turkey came to the rescue.

H. B. Huddleston, 'The Caterpillar Plague: To the Editor', *Bay of Plenty Times*, March 20, 1899.

OPPOSITE
CAROLINA / WOOD DUCK
Aix sponsa

ABOVE LEFT
KHAKI CAMPBELL
Anas platyrhynchos domesticus

Domestic ducks appear to have been first introduced by missionaries, either at the time of Marsden's first visit in 1814 to the Bay of Islands or very shortly after.

F. W. Hutton and J. Drummond, *The Animals of New Zealand: An Account of the Dominion's Air-breathing Vertebrates*, (Whitcombe and Tombs, 1909).

ABOVE RIGHT
MUSCOVY
Cairina moschata

In 1865, Captain Norman liberated six of these birds on Adam's Island, one of the Auckland Islands lying to the south of New Zealand. They failed to establish themselves.

G. M. Thomson, *The Naturalisation of Animals & Plants in New Zealand*, (Cambridge University Press, 1922).

GREYLAG GOOSE
Anser anser

The common goose is particularly interesting as the first species of bird which it was attempted to naturalise in New Zealand. When Captain Cook was at Dusky Sound in March–May 1773 during his second voyage to New Zealand, he liberated some geese. In his *Journal*, he says: 'Having five geese out of those we brought from the Cape of Good Hope, I went with them next morning to Goose Cove (named so on this account) where I left them. I chose this place for two reasons: first, because here are no inhabitants to disturb them; and second, here being the most food. I make no doubt that they will breed, and may in time spread over the whole country and fully answer my intention in leaving them.'

G. M. Thomson, *The Naturalisation of Animals & Plants in New Zealand*, (Cambridge University Press, 1922).

GREYLAG GEESE
Anser anser

A somewhat unusual occupant was to be seen in police kennels in the shape of a fine goose which was discovered by a constable in lower Queen Street at one o'clock this morning. The owner is not requested to call for it as it is to remain in the kennels for a week; it will be fattened and eaten to pay expenses.

Auckland Star, Volume VI, Issue 1777, October 25, 1875.

Mammalia

The first feeling that strikes everyone on coming to New Zealand is its intense want of animal life. Mountains, plains, rivers — mere features without a soul; for you can hardly dignify the miserable ground lark, the wailing weka or the ghoul-like eel with such a title.

Mark P. Stoddart, 'Acclimatisation', *Lyttelton Times*, Volume XXI, Issue 1198, February 18, 1864.

The early settlers of New Zealand found themselves in a land which, as far as regards climate and natural conditions, seemed to them to reproduce many of the best features of the homeland from which they came. They thought with affection and with the glamour of youthful remembrance of the lakes and rivers, the woods and the fields, the hills and the dells of that homeland. They recalled the sport which was forbidden to all but a favoured few, but which they had often longed to share in the game preserves — the deer on the mountains or in the parks, the grouse on the heather-clad hills, the pheasants in the copses and plantations, the hares and partridges in the stubbles and turnip fields, the rabbits in the hedgerows and sandy warrens, and the salmon of forbidden price in their rivers — and there rose up before their vision a land where all these desirable things might be found and enjoyed.

In time, some plants and animals that had been introduced not only established themselves securely, but increased at a rate which upset all calculations. Conditions were produced which had never been anticipated and the introductions became dangerous and expensive pests. Then public measures had to be taken to check the newcomers, and in some cases their natural enemies had to be introduced. This has led to further complications and unexpected results. Those natural enemies, like the things they were meant to check, did not always do what was expected of them; they frequently failed to achieve the purpose for which they were introduced and took to destroying things that should be preserved. Legislation had to be resorted to in order to destroy some introduced things and to protect others.

G. M. Thomson, *The Naturalisation of Animals & Plants in New Zealand*, (Cambridge University Press, 1922).

Livestock

There is not in the world a finer field for experiment than New Zealand. It is a land of rich pasture eminently fitted for the abode of grazing animals. Not one known species of carnivora except the wild cat exists on the island in a state of nature. The rabbit, the hare and various descriptions of antelope and deer would find ample food and shelter on it, and would not interfere with the domestic animals raised and reared under the fostering care of man.

'Acclimatisation', *Otago Witness*, Issue 756, May 26, 1866.

When Rev. Samuel Marsden visited in 1814, he brought with him the mission party, which was established at Kerikeri and Waimate near the Bay of Islands, and the livestock accompanying the party included one entire horse, two mares, one bull and two cows, with a few sheep and poultry. From this date onwards, there is no doubt numerous introductions of plants and animals were made.

From the earliest days of settlement, cattle were run in large numbers on the open country, seldom seeing man and running practically wild. They were gathered together by stockmen at certain times of the year in order to brand the calves, castrate the young bulls, and separate marketable animals. Otherwise, they ran wild, each herd or mob occupying its own particular area of the country.

G. M. Thomson, *The Naturalisation of Animals & Plants in New Zealand*, (Cambridge University Press, 1922).

BLACK HIGHLAND
Bos taurus

ABOVE
WHITE GALLOWAY
Bos taurus

OPPOSITE
HEREFORD CROSS
Bos taurus

The wild cattle of New Zealand are (like the wild pigs) the only domesticated animals which have been running in the unfenced country for several generations back. They are not nearly as abundant today as they were forty or fifty years ago. In those earlier days, most of the cattle on the larger runs were more or less wild. They became greatly excited when they saw a man on foot, for they were mostly accustomed to men on horseback. It is no wonder that when such cattle got into the wild country where they were undisturbed and never saw human beings, they and their progeny quickly became quite wild.

G. M. Thomson, *Wild Life in New Zealand*, (Wellington Government Printer, 1921).

ABOVE
HEREFORD
Bos taurus

OPPOSITE
CHAROLAIS X HEREFORD
Bos taurus

In the vicinity of Waipiropiro Hot Springs roams a wild bull, which is a source of danger to travellers. Yesterday, two horsemen chased the bull. One fired a gun and hit it. The beast charged the other man, named Maguire, overthrowing the horse and rider and killing the horse whilst Maguire was underneath. It then attacked the other man, who fired seven shots into the animal before it gave up the fight and made for the bush. Maguire is not severely hurt.

'Interprovincial', *Ashburton Guardian*, November 22, 1890.

TOP LEFT
TEXEL

TOP RIGHT
SOUTH SUFFOLK

BOTTOM LEFT
GOTLAND

BOTTOM RIGHT
MERINO

OPPOSITE
WILTSHIRE HORN
Ovis aries

In those early days, fences were very rough, and little or no attempt was made to keep the sheep within enclosures. They were therefore allowed to roam freely over the open country and were only mustered at rare intervals for shearing, tailing the lambs, etc. It was inevitable, therefore, that numbers escaped the musterers, especially in high and inaccessible country, and thus wild sheep became very common, especially in the mountain districts of the South Island.

G. M. Thomson, *Wild Life in New Zealand*, (Wellington Government Printer 1921).

OPPOSITE
ARAPAWA
Ovis aries

ABOVE
TEXEL X COOPWORTH
Ovis aries

The first attempt to introduce sheep into New Zealand was made by Captain Cook on his second voyage and was unsuccessful. He brought away two rams and four ewes from the Cape of Good Hope, but by the time the *Resolution* entered Dusky Sound in March 1773, only a ram and a ewe survived, and they were in such a bad state, 'suffering from an inveterate sea-scurvy,' that their teeth were loose, and they could not eat the green food which was given to them. Cook took them on to Queen Charlotte Sound, which was entered on 18th May. In his *Journal*, he says: 'On the 22nd in the morning, the ewe and ram, I had with so much care and trouble brought to this place, were both found dead, occasioned, as was supposed, by eating some poisonous plant. Thus, my hopes of stocking this country with a breed of sheep were blasted in a moment.'

G. M. Thomson, *The Naturalisation of Animals & Plants in New Zealand*, (Cambridge University Press, 1922).

ALPINE *Capra hircus*

TOGGENBURG *Capra hircus*

ANGLO-NUBIAN
Capra hircus

The introduction of goats dates from Captain Cook's second voyage. He says in his *Journal*: 'On 2nd June, 1773, I sent on shore, on the east side of Queen Charlotte Sound, two goats: male and female. The former was something more than a year old, but the latter was much older. She had two fine kids sometime before we arrived in Dusky Bay, which were killed by the cold.'

Mr W. H. Gates of Skippers Canyon writes (April 1916): 'There are a lot of wild goats here, almost within rifle range of my cabin. One sheep farmer gave a shilling per pair of ears and a shilling for each pelt. The male is a rough-looking customer; some have horns fifteen inches in length, and they grow in a slightly spiral form. I think there is a strain of many breeds running through them all. Some have long hair but are not the Angora breed. Some are almost white, but the chief colours are black and white or black and tan.'

G. M. Thomson, *The Naturalisation of Animals & Plants in New Zealand*, (Cambridge University Press, 1922).

ABOVE
BERKSHIRE X LARGE WHITE
Sus domesticus

OPPOSITE
LARGE WHITE
Sus domesticus

Captain Cook let three pigs loose in Queen Charlotte Sound in 1773. He extracted them from the Māori to whom he gave a promise that they would not be killed. He believed that in time, the pigs would stock the whole island. The Māori kept their word, and the navigator's belief was fulfilled. In later years, the 'Captain Cooks', as they were called, afforded a splendid diet for the Māori and early Europeans.

J. Drummond, 'On Introduced Birds,' *Transactions and Proceedings of the Royal Society of New Zealand* 39, (1906).

CAPTAIN COOK BOAR
Sus scrofa

In the case of pigs, it is true that in recent years they have not been an unmixed blessing, owing principally to their weakness for young lamb. But before that delicacy was introduced to their notice, no great sins could be charged against them. Undoubtedly they were the first ploughmen and turned up acres of virgin soil in such a workman-like manner that a little rough bush harrowing made an excellent seedbed for the English grasses that have thriven so well. Piggy did a good turn and destroyed a lot of fern rot and other rubbish; therefore we can safely say Cook's pigs paid for their keep.

A. Gunner, 'The Naturalist: Some Notes on Acclimatisation', *Otago Witness*, Issue 2013, June 14, 1894.

They multiplied astonishingly, and enormous numbers assembled in uninhabited valleys far from the settlements. At Wangapeka Valley, in the Nelson Province, Dr. Hochstetter, in 1860, saw several miles ploughed up by pigs. Their extermination was sometimes contracted out to experienced hunters, and Dr. Hochstetter states that three men in twenty months on an area of 250,000 acres killed no fewer than 25,000 pigs and pledged themselves to kill 15,000 more.

J. Drummond, 'On Introduced Birds,' *Transactions and Proceedings of the Royal Society of New Zealand* 39, (1906).

In some parts of New Zealand, wild pigs are destructive to sheep. I am informed that in North Canterbury, an old boar has been seen to come down from his hill into a paddock in which were a number of ewes, charge into the midst of them and kill two, 'seemingly,' said my informant, 'more out of mischief than for want of food.' Mr W. R. Buen of Kaikōura wrote me in August 1916 'that pigs are very destructive to rabbits, eating the young ones when they take refuge in shallow burrows' and states that 'where pigs are abundant, very few rabbits are to be found'.

G. M. Thomson, *The Naturalisation of Animals & Plants in New Zealand*, (Cambridge University Press, 1922).

BLACK MARE
Equus ferus caballus

To the Editor. Sir, — Is it possible that there is no power in this town of Invercargill that can stop the exercising of trotting horses at top speed in the streets? There are three or four men who make a regular practice of riding such horses as 'hard as ever they can put their feet down', not stopping or easing for crossing or anything; they stand up in the stirrups and with a clutch of the animal's mane, tear along as if they were on a racecourse trying to break a record. Perhaps they would excuse themselves by saying that they were only 'trotting,' but some of these horses can cover a mile in three minutes and are often driven very little short of it in our streets, and at a time when these are busiest. The streets running north and south are mostly used for this illegitimate game, probably because they give the longest, straightest run, but it is time that the impudent practice was sharply stopped.

'Furious Riding', *Southland Times*, Issue 12424, October 14, 1892.

Game

New Zealand should swarm with game. There is no destructive animal or reptile; the climate and soil are perfect; water flows everywhere in profusion; and there is the finest food for everything from jack-snipe to elephant.

Charles Hursthouse, *New Zealand, or, Zealandia, the Britain of the South, Vol.I.*, (Edward Stanford, 1857).

The introduction of game into a country like New Zealand might well be attended with social and even pecuniary benefits which would have tickled a calculator like the late Joseph Hume. We don't go to New Zealand with pick and pan to snatch dear-won nuggets, gulp gallons of rum, and then, rich or ragged, hurry home. We go to the 'Britain of the South' to create an estate.

No man can better deserve, no man can better afford, a day's pastime than a New Zealand colonist. His bow even must sometimes be unbent; and surely 'tis better that he should nerve himself in rural sports 'chasing the red deer and following the roe' than relax himself in city dissipations and the laps of ballet girls.

'Tis a remarkable fact that New Zealand, a country larger than the United Kingdom, possessing a genial climate, abundance of food and cover for all the *Herbivora*, should not possess a single indigenous wild animal. With the exception of two or three small lizards, the kiore (a mouse-rat) is the only four-footed native creature which exists; and even this little king-quadruped of the country is fast disappearing.

Charles Hursthouse, *New Zealand, or, Zealandia, the Britain of the South, Vol.I.*, (Edward Stanford, 1857).

We feel a want for the song of the bird, the consciousness that deer may be feeding on the slopes of our glittering alps, or the salmon or trout splashing in our clear streams.

Mark P. Stoddart, 'Acclimatisation', *Lyttelton Times*, Volume XXI, Issue 1198, February 18, 1864.

FALLOW DEER
Dama dama

The desire to stock the mountain country of New Zealand with large game, so that the Briton's delight in going out and killing something might be satisfied, has led to the introduction of no fewer than ten kinds of deer in addition to other large animals. Of these, four species — red, fallow, white-tailed and Sambar deer — have established themselves in different parts of the country and are included among the animals for which licences to shoot are now issued.

G. M. Thomson, *The Naturalisation of Animals & Plants in New Zealand*, (Cambridge University Press, 1922).

OPPOSITE

WAPITI

Cervus canadensis

Inquiries have been made in regard to moose, elk, wapiti and buffalo from America, Caucasian and Carpathian deer, and Asiatic wapiti and ibex, and it had been decided to obtain elk and buffalo from America. None of the animals could be regarded as dangerous. There were large areas of land unsuited for any other purpose, but if stocked with big game they would be a valuable asset to the colony.

'Town & Country,' *Timaru Herald*, Volume LXXIX, Issue 12190, October 7, 1903.

ABOVE

RED DEER

Cervus elaphus

To the naturalist, it is deplorable that an animal should have been introduced into this sanctuary, which compels men with guns — and probably with dogs — to go in, in order to keep them in check to some extent. Writing to me in August 1918, Mr Hardcastle says: 'Deer increase more rapidly in New Zealand than in the northern hemisphere. Whether there is a larger percentage of calves born, I cannot say, probably there is, considering the conditions here. But the large increase is mainly due to the hinds calving a year earlier.'

G. M. Thomson, *The Naturalisation of Animals & Plants in New Zealand*, (Cambridge University Press, 1922).

ABOVE

ALPINE CHAMOIS
Rupicapra rupicapra

Chamois are increasing in the neighbourhood of Mount Cook. The first were introduced in 1907 when chamois were received as a present from the late Emperor of Austria. By latest accounts, there is now a considerable flock in that mountain region.

G. M. Thomson, *Wild Life in New Zealand*, (Wellington Government Printer, 1921).

OPPOSITE

HIMALAYAN TAHR
Hemitragus jemlahicus

The acclimatisation of deer has been very successful, and a considerable amount of stalking is done in both the North Island and the South. The moose has been let loose on the West Coast of the South Island, and proposals have been made to introduce more big game. The ibex and the tahr are also looked upon as likely beasts for the mountainous regions of the South Island; several of the latter have been introduced by the Tourist Department and have been liberated in the Mount Cook District.

F. W. Hutton and J. Drummond, *The Animals of New Zealand: An Account of the Dominion's Air-breathing Vertebrates*, (Whitcombe and Tombs, 1909).

'Had rabbits not become a nuisance, it is unlikely that weasels and other vermin would have been introduced.'

EUROPEAN RABBIT
Oryctolagus cuniculus

The Acclimatisation Society soon made its existence known and in 1866 received its first gift of birds and beasts from Sir George Grey, who was then governor of the colony. The gift consisted of a pair of white swans, two pairs of black swans and a pair of silver-grey rabbits. It was in this modern way, says our informant, that 'Brer Rabbit' was first introduced to Canterbury. It is not necessary to add that he came to stay.

'The Story of Our Plants and Birds,' *The Press*, Volume LVII, Issue 10840, December 15, 1900.

The introduction of rabbits had a lasting effect on acclimatisation generally. Before their advent, partridges and pheasants had become numerous, but they have entirely disappeared in Otago. In an effort to cope with the rabbits, the country was annually sown with poisoned grain. This had a disastrous effect on both native and imported game. Had rabbits not become a nuisance, it is unlikely that weasels and other vermin would have been introduced. These animals are largely responsible for the decrease in the numbers of native birds and also make the successful introduction of new varieties more difficult. The initial difficulties in getting rabbits introduced, the terrific success that at last crowned the efforts made, the unexpected ruin and destruction which they caused, and the gradual return to normal conditions, makes the history of their introduction one of the most interesting in the annals of acclimatisation.

G. M. Thomson, *The Naturalisation of Animals & Plants in New Zealand*, (Cambridge University Press, 1922).

It appears that the best course to promote the increase of hares in any district would be to destroy the hawks ad lib — having evidently learnt the habits of the hares to such an extent, they will ensure their ultimate annihilation from the country if allowed to go on as they are doing at present.

'Destruction of Hawks', *Lyttelton Times*, Volume LXI, Issue 7240, May 14, 1884.

EUROPEAN HARE
Lepus europaeus

To the Editor of the *Standard*, Sir, — It is a matter which is being discussed just now among some of the settlers, whether the introduction of hares would not be the best means of extirpating the rabbits. The suggestion has been ridiculed by some, but there are others who are inclined to believe that of the two evils, hares are the most desirable. I have heard that one settler in the Lower Valley — Mr McDougall of Porirua — has found by experience that hares and rabbits will not thrive together. According to the increase of hares the rabbits have disappeared, and from what I have heard he attributes the scarcity of rabbits on his grounds to the fact that the hares have overcome them. If this be true it will speedily put an end to the difficulty, for hares do not burrow, and if necessity called for their extirpation also, they could easily be subdued. I should like to hear something upon the subject from others interested.

'Hares v. Rabbits,' *Wairarapa Standard*, Volume XVII, Issue 1443, August 15, 1883.

COMMON BRUSHTAIL POSSUM
Trichosurus vulpecula

Opossums liberated by the Nelson Acclimatisation Society in various parts of the district are said to be increasing. Those set free by the Wellington Society are also doing well. They find excellent food in the native bush as a substitute for the gum forests of Australia.

'Opossums in New Zealand,' *The Press*, January 25, 1896.

A consignment of Tasmanian black opossums has just been liberated at Paekākāriki. A further lot is expected to arrive on Thursday for distribution in different parts of the colony. If the animals thrive in New Zealand as they are expected to do, their acclimatisation will be an item on the assets side of the ledger.

'Miscellaneous,' *New Zealand Mail*, September 8 , 1898.

One of the acclimatisation societies of New Zealand a few years ago introduced the Australian opossum in New Zealand. So rapidly have the opossums increased that they are now a dangerous pest in the native forests and even orchards. In the provincial district of Canterbury, which compared with other parts of New Zealand is somewhat devoid of natural forest lands, the opossums have become so numerous that their consumption of the foliage threatens to end in the utter destruction of the trees.

New Zealand Mail, Issue 988, February 6, 1891.

Predators

Nothing in connection with the naturalisation of wild animals into New Zealand has caused as much heart-burning and controversy as the introduction of these bloodthirsty creatures.

G. M. Thomson, *The Naturalisation of Animals & Plants in New Zealand*, (Cambridge University Press, 1922).

The takahē can hardly fail to become extinct before very long. The country in the neighbourhood of the West Coast Sounds is now swarming with rats, which are playing havoc with the eggs and young of ground-feeding birds, while the ferrets, weasels and other vermin that are being introduced as a counterpoise to bunny will be formidable allies of the rat in this respect. However, I suppose the accumulation of the almighty dollar is a matter of more practical importance than the preservation of the rare and remarkable birds of New Zealand.

C. W. Purnell, 'The Takahē,' *Ashburton Guardian*, Volume VII, Issue 2035, January 12, 1889.

ABOVE

NORWAY RAT

Rattus norvegicus

The HMS *Endeavour* was in Queen Charlotte Sound for some days in January and February 1770, and some of the rats on board were almost certain to find their way ashore. Furneaux arrived in Queen Charlotte Sound in April 1773; Cook in the *Resolution* reaching it in May. Over three years had elapsed between the two visits, and on the second occasion the rats were found to be extraordinarily abundant. The rate of their increase is known to be very great, and it seems to me that the animals met with on the second voyage were the progeny of some which got ashore in 1770.

Meeson describes a plague of these rats in 1884: 'Nelson and Marlborough, in other words, the whole of the extreme northern portion of the South Island of New Zealand, is enduring a perfect invasion. Living rats are sneaking in every corner, scuttling across every path; their dead bodies in various stages of decay strew the roads, fields and gardens, and pollute the wells and streams in all directions. Whatever kills the animals does not succeed in materially diminishing their numbers. Young and succulent crops, as of wheat and peas, are so ravaged as to be unfit for and not worth the trouble of cutting and harvesting. A young farmer the other day killed with a stout stick 200 in a couple of hours in his wheat field.'

G. M. Thomson, *The Naturalisation of Animals & Plants in New Zealand*, (Cambridge University Press, 1922).

OPPOSITE

BLACK RAT

Rattus rattus

The black rat became enormously abundant in the early days of settlement, and used to move about the country in vast armies. The settlers, bush fellers and sawmill hands of fifty to seventy years ago have recorded how invasions of them in countless swarms used to move through their district, climbing everywhere, and eating everything that was of a vegetable nature.

One day, the late Mr R. Paulin and I emerged from the bush in Paterson Inlet when the tide was low . . . we were very much surprised to find the whole beach alive with rats which were feeding on the shellfish and stranded animals that the tide had left exposed. As soon as they saw us they immediately ran for shelter of the bush; they were literally in the hundreds.

G. M. Thomson, *The Naturalisation of Animals & Plants in New Zealand*, (Cambridge University Press, 1922).

OPPOSITE
WEASEL
Mustela nivalis

Another stoat or weasel was brought into Amberley today from Brackenfield. This makes four of these abominable pests which have been seen dead or alive nearby. Complaints have been rife among the farmers of poultry and lambs found dead, with the thug-like brand of the stoat or weasel on them. If this is a sample of what we may expect when the 'natural enemy' of the rabbits becomes thoroughly acclimatised, we have a bad outlook before us.

'Local and General', *Ashburton Guardian*, January 24, 1889.

ABOVE
STOAT
Mustela erminea

My son, G. Stuart Thomson, has given me this note on the destructive action of the stoat. He says: 'At Lee Stream, in the Taieri district, I saw a rabbit paralysed with fright and uttering squeals of terror, and on looking around for the cause observed a stoat ten feet away walking deliberately towards the victim. The rabbit was killed by one bite on the neck. A few weeks ago, a lady informed me that she had seen a somewhat similar occurrence at Brighton, but in this case, the rabbit struggled to the lady for protection and fell trembling at her feet, while the stoat disappeared.'

G. M. Thomson, *The Naturalisation of Animals & Plants in New Zealand*, (Cambridge University Press, 1922).

FERRET
Mustela furo

The Canterbury Society introduced five ferrets in 1867 and an additional one in 1868. They were apparently not liberated, though the progeny was probably sold to private individuals. As rabbits began to increase to an alarming extent, various suggestions were made as to importing what was called 'the natural enemy'.

In this same year, a meeting was held at Masterton to consider the administration of the Rabbit Act and the best means of dealing with the pest. One of the resolutions carried was that 'the introduction of ferrets, stoats and weasels in large numbers is, in the opinion of this meeting, the only means by which the rabbit pest can be successfully put an end to, and every owner of land infested with rabbits should either turn out ferrets in proportion to his acreage or contribute to a fund for the breeding and purchase of ferrets, stoats and weasels to be turned out in the district.'

G. M. Thomson, *The Naturalisation of Animals & Plants in New Zealand*, (Cambridge University Press, 1922).

HEDGEHOG
Erinaceus europaeus

The discovery of a hedgehog at Pleasant Point, remarks the *Timaru Herald*, is a decidedly interesting occurrence, for it has not been publicly known that this animal had been imported, and it certainly is not indigenous. It stands to reason, we think, either that the hedgehog found at Pleasant Point has been brought out as a pet and got loose, or else that hedgehogs have been imported, perhaps accidentally, among straw or packing of some sort, and have bred in the colony. They are very hardy animals, and though almost omnivorous can go for a long time without food. We have often wondered why the acclimatisation societies did not turn their attention to these useful and harmless little creatures. Hedgehogs are the best-known destroyers of small vermin, snakes and black beetles.

'A Hedgeho' [sic], *Wanganui Chronicle*, Volume XXV, Issue 10410, September 24, 1883.

In 1894, the late Mr Peter Cunningham of Merivale, Christchurch, sent a consignment of weka home and got twelve hedgehogs out in exchange. They were placed in a pigeon-house, but got out under the wire-netting and escaped. For years nothing was heard of them, but they gradually increased, and are now extraordinarily abundant. Mr Edgar F. Stead of Riccarton says (March, 1916): 'If I hunted through my garden with my dog I could get a dozen now, and I frequently kill them . . . They are extraordinarily destructive to chickens, their depredations being readily identified by the fact that they eat their victim's stomach first, whereas a cat eats the breast first, and rats and weasels go for the head and neck. Once a hedgehog starts eating chickens, he will go on until caught or the supply runs out. I know of many cases when a trap set and baited with the remains of a chicken has caught the marauding hedgehog.

G. M. Thomson, *The Naturalisation of Animals & Plants in New Zealand*, (Cambridge University Press, 1922).

ABYSSINIAN
Felis catus

Mr Robert Scott, M.P. for Otago Central . . . says: 'The wild cat was no doubt the descendant of the shepherds' and miners' tame cat. The predominating colour was grey-striped, or tiger-striped as some people called them; occasionally yellow, and rarely black or black and white. The time I write of was the seventies, say from 1870 on to the time when poisoning the rabbits with phosphorised grain came in. The cats, though not numerous, were fairly common, especially in districts where cover, such as fern and scrub, was plentiful. They grew to an immense size and were game to the last if attacked; in fact, no dog would tackle one single-handed.'

Wild cats have been found in New Zealand from the early days of settlement, though for a long time, they never strayed very far from the abodes of men. But after rabbits began to increase in many parts at such a rate as to reduce the sheep-carrying capacity of the country, sheep farmers began to purchase cats in towns. These were taken out to the backcountry, turned out and fed for a time till they were established.

G. M. Thomson, *The Naturalisation of Animals & Plants in New Zealand*, (Cambridge University Press, 1922).

ABOVE
HUNTAWAY
Canis lupus familiaris

OPPOSITE
HEADING DOG
Canis lupis familiaris

Forster, in his account of the second voyage, 1773, says: 'A good many dogs were observed in their canoes, which they seemed very fond of, and kept tied with a string around their middle; they were of a rough, long-haired sort, with pricked ears, and much resembled the common shepherd's cur or Count Buffon's chien de berger. They were of different colours, some quite black and others perfectly white. The food which these dogs receive is fish, or the same as their masters live on, who afterwards eat their flesh and employ the fur in various ornaments and dresses'.

Later on in the same journal, he says: 'The officer ordered their black dog to be killed and sent to the captain . . . therefore we dined for the first time on a leg of it roasted, which tasted so exactly like mutton that it was absolutely indistinguishable.'

G. M. Thomson, *The Naturalisation of Animals & Plants in New Zealand*, (Cambridge University Press, 1922).

SCOTTISH DEERHOUND
Canis lupus familiaris

It is not generally known that around Otamatea and Wairoa, the bush is infested with packs of wild dogs, as ferocious but more daring than wolves. These dogs hunt in packs of three to six at a time. They are strong, gaunt, large animals, and dangerous when met by man alone. Not long since, a man, when travelling from one settlement to another through the forest, was attacked by three of these animals at dusk and only saved himself by climbing into a tree where he was kept prisoner until the next day. The extensive district to which these packs roam was once well stocked with wild pigs, but most of these have fallen victim to the dogs, and since their supply of food has failed, the dogs have ventured after dark to the neighbourhood of Māori settlements and the homesteads of European settlers in quest of prey.

'Bush Dogs in the North', *Taranaki Herald*, Volume XV, Issue 748, December 1, 1866.

Dieffenbach, writing nearly seventy years after Cook's visit, remarks that: 'the native dog was formerly considered a dainty, and great numbers of them were eaten; but the breed has undergone an almost complete mixture with the European and their use as an article of food has been discontinued, as the European dogs are said by the Māori to be perfectly unpalatable.'

G. M. Thomson, *The Naturalisation of Animals & Plants in New Zealand*, (Cambridge University Press, 1922).

The Acclimatisation Society has, of course, made experiments which have failed, but the dream of its founders has been realised, and bird and animal life now enliven the erstwhile solitude with their cheerful presence.

The Story of Our Plants and Birds,' *The Press*, Volume LVII, Issue 10840, December 15, 1900.

Chronology of Introductions

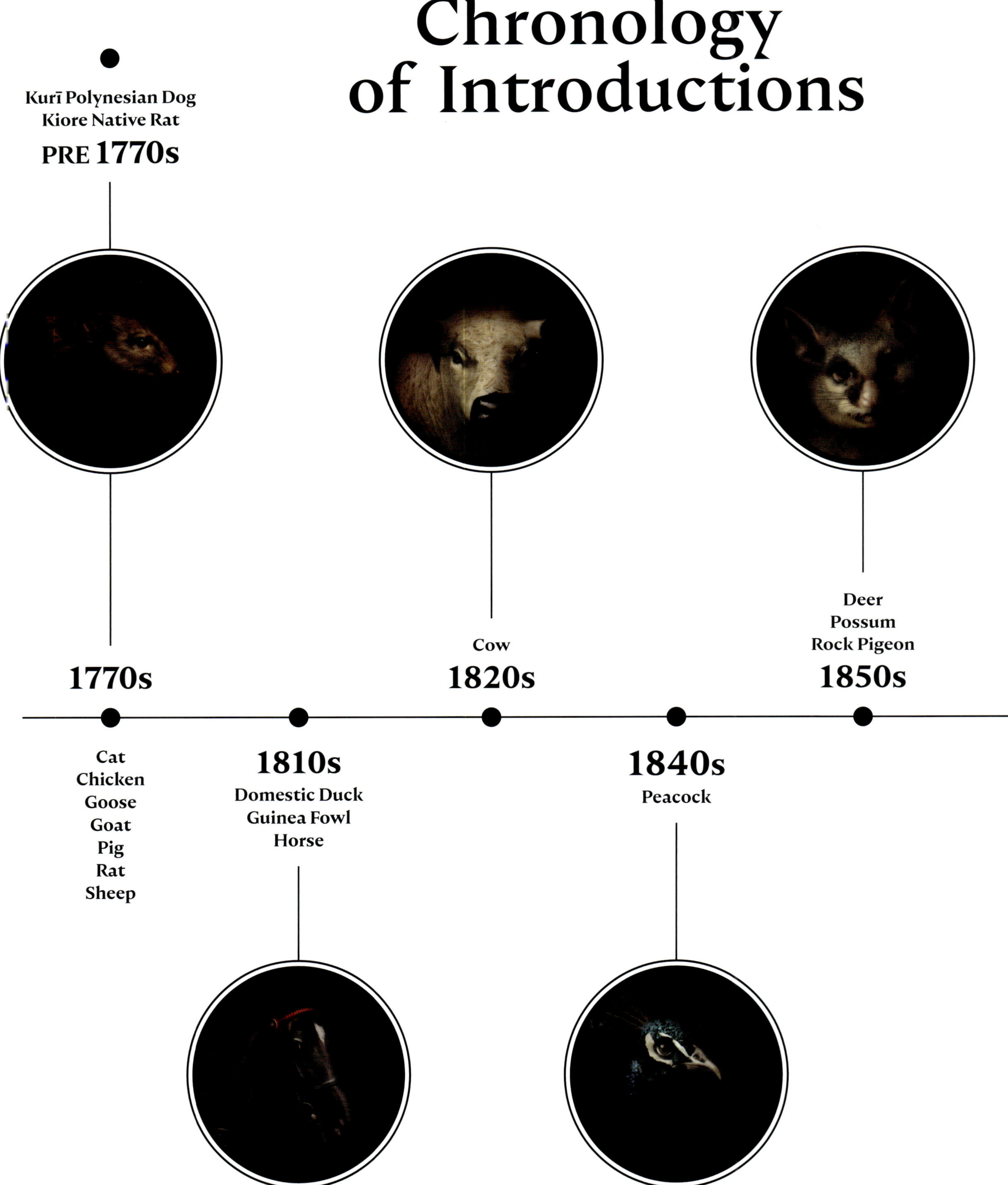

Australian Magpie
Black Swan
Chaffinch
Common Pheasant
Common Starling
Dunnock
Eastern Rosella
Eurasian Blackbird
European Rabbit
Fallow Deer
Ferret
Goldfinch
House Sparrow
Laughing Kookaburra
Mallard Duck
Mute Swan
Quail
Redpoll
Song Thrush
Turkey

1860s

Common Myna
Hare
Hedgehog
Partridge
Rook
Wallaby

1870s

1880s

Stoat
Weasel

Alpine Chamois
Canada Goose
Himalayan Tahr
Little Owl

1900s

Part 2

Native & Endemic

The bush is frequented by a variety of birds — the kākāpō, crow and saddleback, besides others of less importance. The kākāpō is a species of parrot about the size of a small fowl; the crow is about twice the size of a tūī, of a bluish tint with beautiful red gills; the saddleback is of a bluish tone with the shape of a pack saddle beautifully marked in brown upon its back — hence its name. There is also a small bird about the size of two thumbs called the rock wren, which finds its home under the steep bluffs among the fragments of broken rock. All these birds are sought after to furnish our museums.

'A Trip to the Tableland: Range of Birds and Taxidermy,' *Nelson Evening Mail*, Volume XXXII, Issue 208, September 9, 1898.

Aves

'All the more interesting of these forms are passing away. Not a few species have already been exterminated, many more on the borderland, so to speak, of final extinction, and some even of the commonest birds of five and twenty years ago have become so scarce that it is difficult to know where to look for them.'

Only a few weeks ago, I received a letter from a leading scientist on the Continent, in which he says: 'I have always regarded the avifauna of New Zealand as the most interesting in the world. I was therefore delighted to read your spirited request to the authorities to do something, if only to save a remnant of it, and failing that, to get representative collections in all the local museums before it is too late. The cost of this great and important work would be comparatively trifling – less, in fact, than the expense of equipping a small gunboat for active service in your waters.' And this estimate of the New Zealand avifauna is no mere figure of speech. It contains so many anomalous genera – so many types of ancient forms or connecting links with a fauna of the past – that its study is of the highest interest to the philosophic naturalist. And what is happening at this moment? Why, all the more interesting of these forms are passing away. Not a few species have already been exterminated, many more are on the borderland, so to speak, of final extinction, and some even of the commonest birds of five and twenty years ago have become so scarce that it is difficult to know where to look for them. The saddest part of it is that it seems hopeless now to arrest the evil. The Government has at last been roused to take action, and by extending the provisions of the Wild Birds Protection Act, it shields, to some extent, species that were being indiscriminately destroyed, but those more in need of protection have become exposed to the deadliest of natural enemies; the introduction of stoats, weasels and ferrets – bloodthirsty animals that are now swarming over every part of the country, and I defy all attempts to check their increase. The intention, of course, was to find a new remedy for the wild rabbit nuisance, but it is notorious that these marauders will not take fur when they can get feather. From all parts of both islands, I receive intelligence of the ravages of these animals, and being now thoroughly established in the country, it will be impossible ever to eradicate them. Let it be understood, therefore, that many of our indigenous species are doomed, and every true naturalist must grieve over it as an irreparable loss to science.

Professor Newton of Cambridge, in an address which he gave to the British Association in 1887 referring to the rapid extinction, even at that time, of many New Zealand birds, said, 'I would ask you to bear in mind that these indigenous species are, with scarcely an exception, peculiar to that country, and from every scientific point of view, of the most instructive character. They supply a link with the past, which once lost can never be recovered . . . The forms that we are allowing to be killed off, being almost without exception ancient forms, are just those that will teach us more of the way in which life has appeared on the globe than any other recent forms.'

Walter L. Buller, 'The Vanishing Forms of Bird-Life in New Zealand', *The Press*, Volume LIV, Issue 9622, January 11, 1897.

KERERŪ
NEW ZEALAND PIGEON
Hemiphaga novaeseelandiae
Not threatened

Woodland

It is yet possible to reach some secluded spots where the hanging branches of virgin forests exhibit lovely forms and hues of glorious foliage in all their pristine beauty, still unscathed by fire or bushman's axe; where birds still flutter and carol through revolving seasons of a golden age; where the murderous guns and the stealthy cat are alike unknown; where the confident curiosity of the birds surprises the human trespasser and teaches him, in the plainest language, the story of the changes effected by the savage barbarity of man.

'On the Birds of New Zealand,' *Transactions and Proceedings of the Royal Society of New Zealand* 3, (1870): 59.

When civilisation had fairly established itself, bringing many species of its domestic animals and several species of its domestic pests and vermin, there began a short, sharp and bitter struggle between the new fauna and the old one, which had possessed this country for ages.

The result was never in doubt. The members of the old fauna, which may be regarded as aristocrats of the animal kingdom, had absolutely no chance against the shrewd, vulgar, hard-headed, cunning, practical, greedy and ferocious invaders, who were inured to hardship and had walked hand in hand with adversity through many generations. The incident was an especially dramatic one with respect to the avifauna.

J. Drummond, *Our Feathered Immigrants: Evidence For and Against Introduced Birds in New Zealand, Together with Notes on the Native Avifauna*, (Wellington Government Printer, 1907).

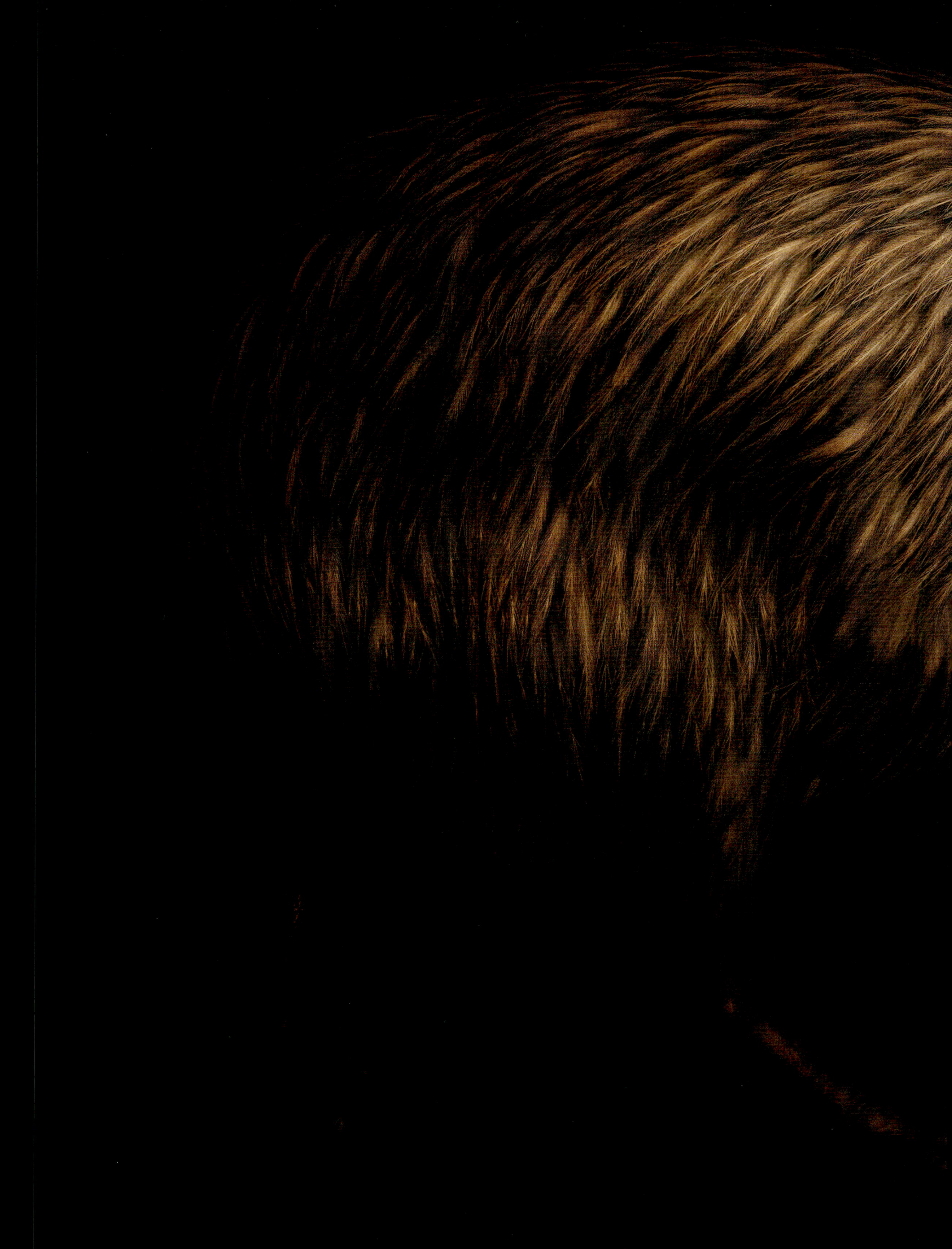

OPPOSITE

RUROA | GREAT SPOTTED KIWI

Apteryx maxima | **Nationally vulnerable**

Dogs readily follow the scent of *Apteryx*. Very large numbers of the birds have been needlessly destroyed by hunters. Bushmen, it is stated, do not dislike the taste of the kiwi, and the meat, though coarse, has a gamey flavour. The eggs afford excellent food. At one time, a great demand sprang up for kiwi skins, which were used for making muffs, and the hunting of the birds became almost a vocation. One kiwi hunter told Mr Potts that up to the close of 1871, he had killed about 2200 specimens of *Apteryx oweni* and *Apteryx australis*.

F. W. Hutton and J. Drummond, *The Animals of New Zealand: An Account of the Dominion's Air-breathing Vertebrates*, (Whitcombe and Tombs, 1909).

PREVIOUS

KIWI-NUI
NORTH ISLAND BROWN KIWI

Apteryx mantelli | **Not threatened**

The kiwi prefers a worm above all other foods. First comes one and then another kiwi, in search of worms, their heads always on one side, with an ear turned to the ground, listening for the creeping of the worm beneath the soil. As soon as one of them hears one, down goes his beak, right to the worm, which it brings up to eat. The noise is like that made by the hand of a watch but rather louder. That is what the kiwi listens for. On account of this habit, the hunter carefully prepares little pieces of wood, which are tied to the dogs' necks so that they may rattle as the animals move. Hearing this, the kiwi thinks that it is a worm and stops to listen. While it is doing this, the dogs are able to approach, and by the time it starts to run, the dogs are baiting it. The men then advance with their torches, which are burning, and consequently, the birds cannot see and are caught and killed. The kiwi are never hunted by daylight.

F. W. Hutton and J. Drummond, *The Animals of New Zealand: An Account of the Dominion's Air-breathing Vertebrates*, (Whitcombe and Tombs, 1909).

TOP

KAKARUWAI
SOUTH ISLAND ROBIN

Petroica australis **| Declining**

The South Island robins may often be seen in the bush. They enter tents, hop about inside, and sometimes partake of meals on the table. They are active little fellows. They appear at the break of day and retire when darkness comes on. They have been known to sit on a man's body as he lay on the ground, hopping about and picking at his watch chain or at any other object that attracted their attention.

F. W. Hutton and J. Drummond, *The Animals of New Zealand: An Account of the Dominions Air-breathing Vertebrates*, (Whitcombe and Tombs, 1909).

BOTTOM

TĪTITIPOUNAMU | RIFLEMAN

Acanthisitta chloris **| Not threatened**

The rifleman inhabits generally the sub-alpine forests of both islands as well as Great Barrier and Little Barrier Islands, but very little is known of its habits. It forms its nest in hollow trees, and it is stated that a pair of these little birds once took up their residence in the skull of a horse.

F. W. Hutton and J. Drummond, *The Animals of New Zealand: An Account of the Dominion's Air-breathing Vertebrates*, (Whitcombe and Tombs, 1909).

TOP

PŌPOKOTEA | WHITEHEAD

Mohoua albicilla | Not threatened

The whitehead is one of the native birds whose doom has been announced, but it is satisfactory to learn that it is plentiful in both the Little Barrier and Kāpiti sanctuaries and, according to recent information, in many places on the mainland of the North Island. It affords one of the pleasantest scenes of the forest. When all other birds are absent or silent, it comes along and enlivens the forest with its busy twitter and it's extraordinarily quick movements.

F. W. Hutton and J. Drummond, *The Animals of New Zealand: An Account of the Dominion's Air-breathing Vertebrates*, (Whitcombe and Tombs, 1909).

BOTTOM

TAUHOU
SILVEREYE / WHITE-EYE

Zosterops lateralis | Not threatened

Before 1856, the white-eye was unknown to either Europeans or Māori in New Zealand, and, when it came here, in that year, the latter named it 'tauhou', which means 'stranger'. By 1861 it had spread all over the South Island, and the southern parts of the North Island, though it did not reach Auckland until 1865. Its presence here shows that the passage across the Tasman Sea is possible even for some small land birds. The distance, as the crow flies, is about a thousand miles; and it must take a bird, flying at ordinary speed, from twenty-four to thirty-six hours to accomplish the journey.

F. W. Hutton and J. Drummond, *The Animals of New Zealand: An Account of the Dominion's Air-breathing Vertebrates*, (Whitcombe and Tombs, 1909).

PĪWAKAWAKA
FANTAIL
Rhipidura fuliginosa
Not threatened

When a dry summer comes, the Auckland youths swarm to the Waitākere Ranges with tent and gun, and everything with life is shot down. His unquenched fire spreads and often burns for weeks. After one such raid, you can pick up scores of beautiful, harmless fantails and other small birds on the roads; they are shot for pure cussedness, for a benefit to the destroyer they are none.

J. McElwain, 'Destruction of Birds in Waitākere Reserve,' *Auckland Star*, Volume XXVI, Issue 107, May 6, 1895.

KŌTARE
SACRED KINGFISHER
Todiramphus sanctus
Not threatened

The New Zealand kingfisher does not live by fish alone. He has an insatiable appetite and will eat mice, bees, beetles and other insects, *Crustacea*, and even young birds and chickens. The omnivorous little rascal was accused, some years ago, of catching trout in the Acclimatisation Society's Gardens in Christchurch, until the artificial races were netted in to keep him off, and also of killing Californian quails in the Auckland Gardens.

F. W. Hutton and J. Drummond, *The Animals of New Zealand: An Account of the Dominion's Air-breathing Vertebrates*, (Whitcombe and Tombs, 1909).

ABOVE
PĪPĪWHARAUROA
SHINING CUCKOO
Chrysococcyx lucidus | Not threatened

OPPOSITE
KOEKOEĀ
LONG-TAILED CUCKOO
Eudynamys taitensis | Nationally vulnerable

The little grey warbler is generally selected by the cuckoos as a victim for their parasitical habits. Instances are recorded, however, where the nests of other birds have been appropriated. In Otago, the grey warbler, the South Island tomtit, and the white-eye are generally made the foster parents.

In New Zealand, the little intruder probably follows the tactics adopted by these birds in England, where the young cuckoo lifts the other young birds on its back and tumbles them overboard.

F. W. Hutton and J. Drummond, *The Animals of New Zealand: An Account of the Dominion's Air-breathing Vertebrates*, (Whitcombe and Tombs, 1909).

KORIMAKO | BELLBIRD

Anthornis melanura **| Not threatened**

Over a hundred years ago, Captain Cook described the delight he experienced when this bird's splendid song first fell upon his ears. It was in Queen Charlotte Sound when the ship *Navigator* was about a quarter of a mile from shore. 'And in the morning,' he says, 'we were awakened by the singing of the birds. The number was incredible, and they seemed to strain their throats in emulation of each other. This wild melody was infinitely superior to any that we have ever heard of the same kind: it seemed to me like small bells, most exquisitely tuned, and perhaps the distance and the water between might be no small advantage to the sound.'

The decrease in the number of the bellbirds is attributed to bushfires, especially in the North Island, and to the depredations of cats, rats and honeybees.

F. W. Hutton and J. Drummond, *The Animals of New Zealand: An Account of the Dominion's Air-breathing Vertebrates*, (Whitcombe and Tombs, 1909).

OPPOSITE

TĪEKE | NORTH ISLAND SADDLEBACK

Philesturnus rufusater | Relict

ABOVE

NORTH ISLAND KŌKAKO

Callaeas wilsoni | Nationally increasing

This bird is becoming very rare where formerly it abounded. During a recent expedition into the Tararua Ranges extending over three days, I met with only one, a fine adult male, which I had no difficulty in shooting as the bird is not shy. It made a beautiful cabinet specimen, although soon after the death the rich mazarine blue faded out of the wattles, and they became black as the skin dried.

Walter L. Buller, 'Further Notes on the Birds of New Zealand,' *Transactions and Proceedings of the Royal Society of New Zealand* 25, (1892): 63.

TŪĪ

Prosthemadera novaeseelandiae | **Not threatened**

In the winter the tūī is very fat, and consequently, like the pigeon and other birds worth shooting for food, will probably disappear as the country gets settled.

'Notes from the Bush,' *Wellington Independent*, Volume XIX, Issue 2166, February 11, 1865.

Caught and caged, it is still the merry ventriloquist and mocks, cocks and cats and attempts the baby. To add to its merits, it becomes very fine eating in the season of poroporo berries.

F. W. Hutton and J. Drummond, *The Animals of New Zealand: An Account of the Dominion's Air-breathing Vertebrates*, (Whitcombe and Tombs, 1909).

KERERŪ | NEW ZEALAND PIGEON

Hemiphaga novaeseelandiae **| Not threatened**

Our wood pigeon — the largest and most handsome of the fruit-eating pigeons, to which it belongs — was diminishing so rapidly all over the country that the Legislature very wisely passed a special Act, making the year 1896 and every sixth year thereafter a strictly closed season for this bird. It is, of course, the apterous and ground-feeding birds that suffer most from the depredations of stoats, weasels and polecats, but we have recently had undoubted proof that in the Upper Whanganui and Tongariro country — where these animals abound in great numbers — the pigeons, which habitually roost in low trees, suffer severely from them. The marauders, which are nocturnal in their habits, climb the trees at night and attack the pigeons in their sleep for the purpose of sucking their throats. The numerous bodies of pigeons found lying in the woods punctured in the neck sufficiently attest the fact.

'The Vanishing Forms of Bird-Life in New Zealand,' *The Press*, Volume LIV, Issue 9622, January 11, 1897.

The beautiful plumage of the native wood pigeon gives it rank among the most handsome birds belonging to this dominion. Unfortunately for itself, its flesh is very palatable, and pigeon pie is a dainty dish that is much relished.

F. W. Hutton and J. Drummond, *The Animals of New Zealand: An Account of the Dominion's Air-breathing Vertebrates*, (Whitcombe and Tombs, 1909).

BUFF WEKA
Gallirallus australis hectori
At risk / relict

With weka, thieving is a disease. They are confirmed kleptomaniacs. They steal not only from other birds but also from human beings and commit the crime, apparently, merely for the sake of stealing, as the articles they sometimes appropriate can be of no earthly use to them. Anything, from ducks' eggs to spoons, pipes, and pannikins, is good enough to be carried off.

In speaking of the good qualities of the weka, Mr W. W. Smith denounces the 'ignorant and mistaken prejudice' shown towards this remarkable bird. He finds that too much cannot be said in its favour and believes that all prejudice would be overcome if proper attention were given to its habits, as 'the mere destruction of a few eggs in or near the poultry yard, or the disturbing of a few pheasants in reserves, may be overlooked, wherein it is remembered that the weka renders inestimable services in destroying vermin.' He adds that its struggle for existence is greater than that of any other native bird. This is owing to thousands perishing annually in the fires that swept over large areas of tussock lands, to merciless destruction by dog and gun, and, above all, to stoats and weasels.

F. W. Hutton and J. Drummond, *The Animals of New Zealand: An Account of the Dominion's Air-breathing Vertebrates*, (Whitcombe and Tombs, 1909).

ABOVE
KĀKĀRIKI KARAKA
ORANGE-FRONTED PARAKEET
Cyanoramphus malherbi
Nationally critical

OPPOSITE
KĀKĀRIKI
RED-CROWNED PARAKEET
Cyanoramphus novaezelandiae
Relict

Writing in regard to Otago, in 1877, Mr E. Gillies says: 'To say that they are continually to be seen in flocks of hundreds gives a very faint idea of their extreme prevalence. Settlers whose cultivations were in the bush always had the greatest difficulty in saving crops of wheat. These lively, roguish little birds defied all scarecrows and even shooting was found to be an endless and expensive job, for, though a few might be killed at a blow, the flock just rose and settled down again immediately a few yards off. So rare and scarce have they become now that country settlers near the bush have quite a warm side to the little green parakeets and often make household pets of them.

F. W. Hutton and J. Drummond, *The Animals of New Zealand: An Account of the Dominion's Air-breathing Vertebrates*, (Whitcombe and Tombs, 1909).

KEA
Nestor notabilis
Nationally endangered

It is well known that, to settlers in many parts of the outlying districts of Otago, Southland and Canterbury, the kea has become a source of great anxiety because of the evil habit it has acquired of attacking sheep and eating the kidney fat. Alighting on the back of a sheep, the bird digs its beak through the skin and flesh until it secures the morsel and makes its cruel repast. Very startling figures showing the destruction caused by these birds have been published. Although some of the statements seem to be exaggerated, there is no doubt that the settlers' flocks have suffered heavily. The local governing bodies and the Government still pay for the head of kea, the County Council in some instances giving half a crown each, and the Government granting a subsidy of sixpence a head. Besides this, some settlers have offered as much as ten shillings a head. This system of outlawry has considerably thinned the numbers of the kea. Some settlers slay their hundreds, others their thousands.

F. W. Hutton and J. Drummond, *The Animals of New Zealand: An Account of the Dominion's Air-breathing Vertebrates*, (Whitcombe and Tombs, 1909).

NORTH ISLAND KĀKĀ
Nestor meridionalis
Recovering

During the winter, the kākā often entered the house for food, making themselves thoroughly at home and even roosting on the cross-beams in the kitchen on especially inclement nights. Two of them, in particular, soon learnt how to open the door of the dairy, which they were fond of getting into, in order to regale themselves on cream and butter.

Mr Travers had several of these birds sitting on the eaves of the house in the evening, waiting to be fed and coming readily to receive from the hand pieces of bread thickly spread with butter and strewn with sugar. But they rarely accepted any of the bread, dropping it as soon as they had cleared off the butter and sugar. When anxious to get into the house, they took up positions on the windowsills and beat at the window with their beaks until they were admitted. In some instances, their familiarity degenerated into such gross impudence that Mr Travers's manager was obliged to kill them in order to put an end to their mischief.

F. W. Hutton and J. Drummond, *The Animals of New Zealand: An Account of the Dominion's Air-breathing Vertebrates*, (Whitcombe and Tombs, 1909).

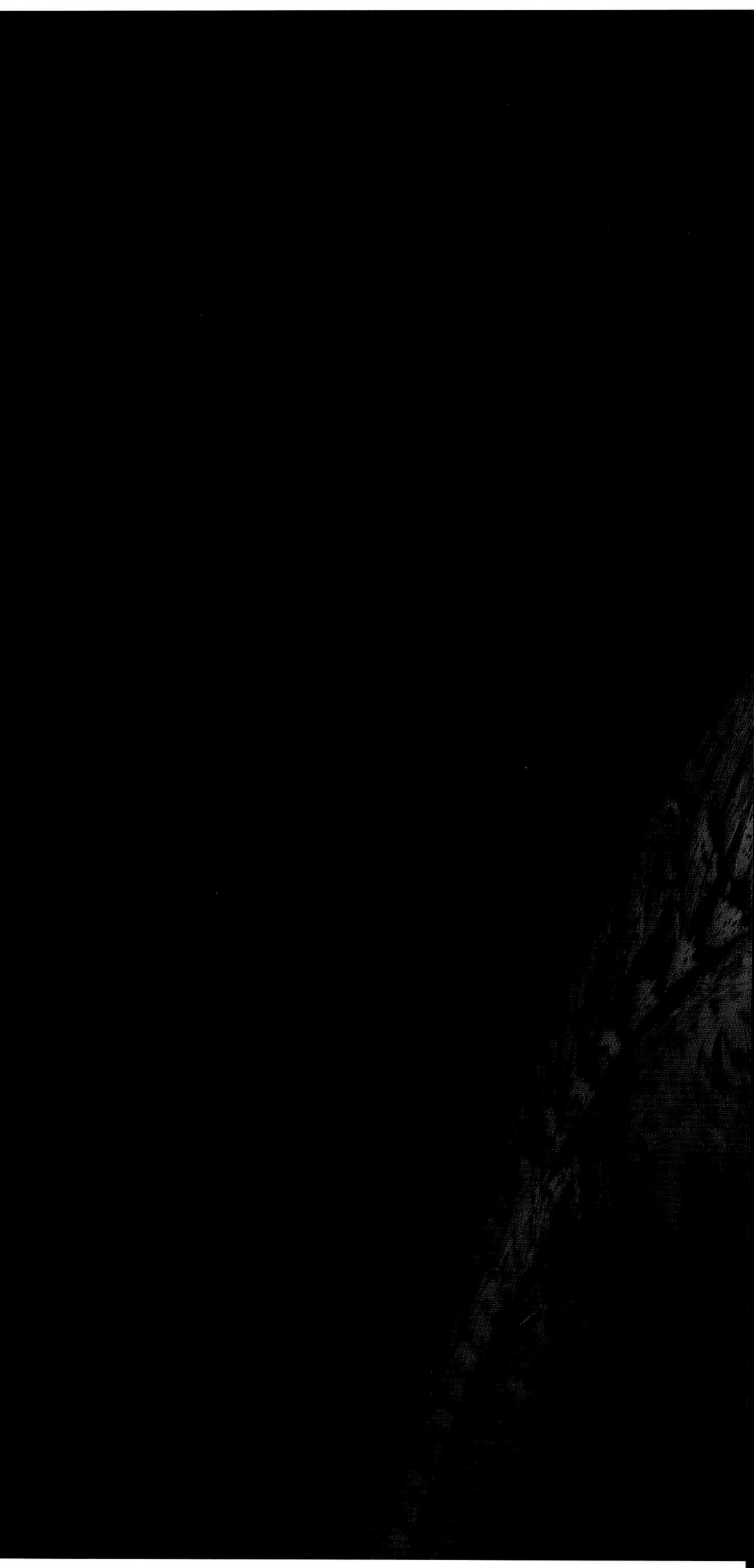

KĀKĀPŌ
Strigops habroptilus **| Nationally critical**

They are strictly nocturnal and repose during the day in underground holes or in hollow logs, so that they are specially accessible to wild dogs and cats and to the introduced marauders of which I have spoken. The kākāpō was so numerous in Haast's time that, for many months together, it formed the staple item in the bill of fare of his exploring party.

Walter L. Buller, 'The Vanishing Forms of Bird-Life in New Zealand', *The Press*, Volume LIV, Issue 9622, January 11, 1897.

The first time we saw the kākāpō by daylight was on an afternoon. The bird was sitting on a stump in an open part of the forest, not far from the Haast River. On our approach, it quickly disappeared but was soon caught by the dog.

F. W. Hutton and J. Drummond, *The Animals of New Zealand: An Account of the Dominion's Air-breathing Vertebrates*, (Whitcombe and Tombs, 1909).

RURU | MOREPORK
Ninox novaeseelandiae **| Not threatened**

The morepork is a bird of the night, and the granivorous and insectivorous birds introduced by the Society seek their food by day, so it is difficult to see how they can come in collision. To attack a brood of nestlings is much more in the line of business of a rat than of a morepork; and as the destruction of one morepork will probably save the lives of some hundreds of rats, the shilling proposed to be paid for the moreporks' heads is hardly likely to prove a good investment. If moreporks are destructive, the rats on which they prey are ten-fold more so, in a variety of ways.

Agricola, 'The Morepork', *Daily Southern Cross*, Volume XXIV, Issue 3382, May 19, 1868.

KĀREAREA | NEW ZEALAND FALCON

Falco novaeseelandiae | Recovering

The quail-hawk, when in pursuit of its prey, is courageous and persevering. It has made a bad reputation for itself as a daring marauder. It will swoop down on a farmyard, strike a fowl, and cling to it until knocked over with a stick. It seems to be quite insensible to danger, darting into houses and following its prey into inner rooms. This bold mountaineer especially delights in capturing domestic pigeons, which afford a tasty meal. Having secured its prey, it takes it to a sheltered spot, places it down rather carefully, and devours it at once.

F. W. Hutton and J. Drummond, *The Animals of New Zealand: An Account of the Dominion's Air-breathing Vertebrates*, (Whitcombe and Tombs, 1909).

KĀHU | SWAMP HARRIER

Circus approximans | **Not threatened**

The sheep-farmer protects [the hawk] because it is more or less destructive to rabbits which are his great pest, though it is really too slow in its movements to keep down such an active and wary creature as the rabbit . . . Wherever game fanciers prevailed, the bird was mercilessly destroyed, and rewards were offered for heads, as it certainly is a most active agent in keeping down pheasants, quail, wild ducks and other imported game. Where farmers grew grain crops the hawk was protected as an antidote to the small bird pest; but where poultry were largely kept, it was ruthlessly trapped and destroyed on account of its predilection for chickens. Altogether the record is a very mixed one, but on the whole the bird seems to have more than held its own, and is very common throughout New Zealand.

G. M. Thomson, *The Naturalisation of Animals & Plants in New Zealand*, (Cambridge University Press, 1922).

Lake & Wetland

'For it is in these salt marshes that these birds get the food they love, and grow to that plump perfection which renders them so delicious.'

Our native waterfowl, unlike those of Britain, the *Mail* said, 'are not migratory but live with us all the year round. The grey and paradise ducks, spoonbill ducks, and black and brown teal, all used to go inland to breed. The young were reared in the creeks, running in every part of the Peninsula, and, in the autumn, when they grew strong enough on the wing, the parents led them to Lakes Ellesmere and Forsyth, and other lagoons and marshes bordering on the sea, for it is in these salt marshes that these birds get the food they love, and grow to that plump perfection which renders them so delicious.'

F. W. Hutton and J. Drummond, *The Animals of New Zealand: An Account of the Dominion's Air-breathing Vertebrates*, (Whitcombe and Tombs, 1909).

ABOVE

WEWEIA | NEW ZEALAND DABCHICK

Poliocephalus rufopectus | Nationally increasing

OPPOSITE

PŪTEKETEKE | AUSTRALASIAN CRESTED GREBE

Podiceps cristatus | Nationally vulnerable

To the Editor. Sir, — As this Society is anxious to protect all native game, I think it desirable that the public should be made aware that the taking of the crested grebe, as reported in your issue of yesterday, was an illegal act, and the persons so offending were liable to a heavy penalty. I may also mention that the tūī, huia, white heron and the bittern are absolutely protected.

H. A. Bruce, 'Protected Birds', *Lyttelton Times*, Volume XCIII, Issue 10616, March 27, 1895.

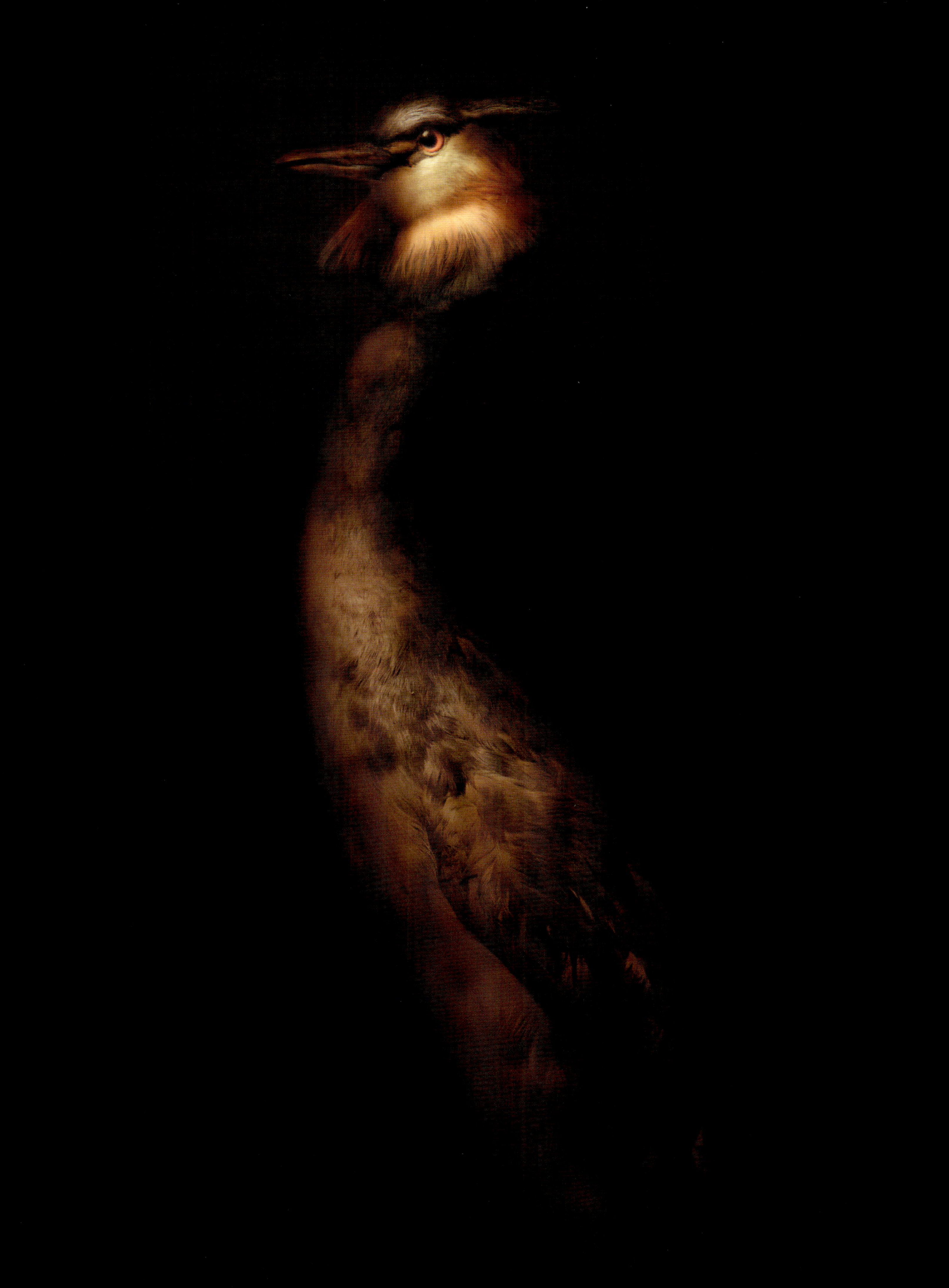

ABOVE
PŪTANGITANGI | PARADISE SHELDUCK
Tadorna variegata | Not threatened

TOP RIGHT
KURUWHENGI | AUSTRALASIAN SHOVELER
Spatula rhynchotis | Not threatened

BOTTOM RIGHT
PĀTEKE | BROWN TEAL
Anas chlorotis | Nationally increasing

In many cases, the shooting season begins too early. In late summers and fine open winters the paradise duck rears a second brood, or at least breeds late in the autumn, and when sportsmen come round with breechloaders and pour a perfect fusillade of leaden hail into a flock — many young ducks, which are not strong of wing — sometimes barely able to fly — come to grief at once. The spirit of destructiveness in Englishmen is so strong that it is no wonder that the Frenchman witheringly remarked that when they had nothing to do they would go out and 'kill something'.

'The Paradise Duck', *Otago Witness*, Issue 1921, December 11, 1890.

RIGHT AND PAGES 170–171

PĀPANGO
NEW ZEALAND SCAUP
Aythya novaeseelandiae
Not threatened

The black teal got tame in one day, and would not leave the pool while the canoe drifted slowly round it, and seemingly we could have caught them on our hooks if we had tried. Probably we had more pleasure in their company and confidence than we should have had in killing and eating them, for then we should have had the plucking and cooking to do, and we had excellent tinned meat that needed none of that bother. We caught two eels, which we shared with the dog, and they were very acceptable.

R. Henry, 'Exportation from Dusky Sound,' *Otago Daily Times*, Issue 10613, March 7, 1896.

ABOVE

TĒTĒ-MOROITI | GREY TEAL

Anas gracilis **| Not threatened**

The shooting season for native game opened on Monday, and during the ensuing holidays hundreds of sportsmen will endeavour by all the means at their command to still further reduce the fast-diminishing numbers of wild ducks and such other native birds as the Government and the Acclimatisation Society in their magnanimity choose to brand for the slaughter. Time was when wild ducks sat in moving masses on our lakes — thousands upon thousands, covering acres at a time — and every sheltered nook by river and stream was alive with bird life. But in the days when acclimatisation societies were not, and every man was a law unto himself, the wild ducks were shot down by the thousands.

'Our Native Game', *Otago Witness*, Issue 2455, 3 April, 1901.

OPPOSITE

WHIO | BLUE DUCK

Hymenolaimus malacorhynchos **| Nationally vulnerable**

The river courses are frequented by a good many birds of the different varieties of duck, teal and divers. Far up in some of the rivers, a blue duck with a yellow bill occurs. It is called the whio and has a melancholy whistling cry. When pursued, it does not fly far and is easily overtaken and shot.

'Notes From the Bush', *Wellington Independent*, Volume XIX, Issue 2166, February 11, 1865.

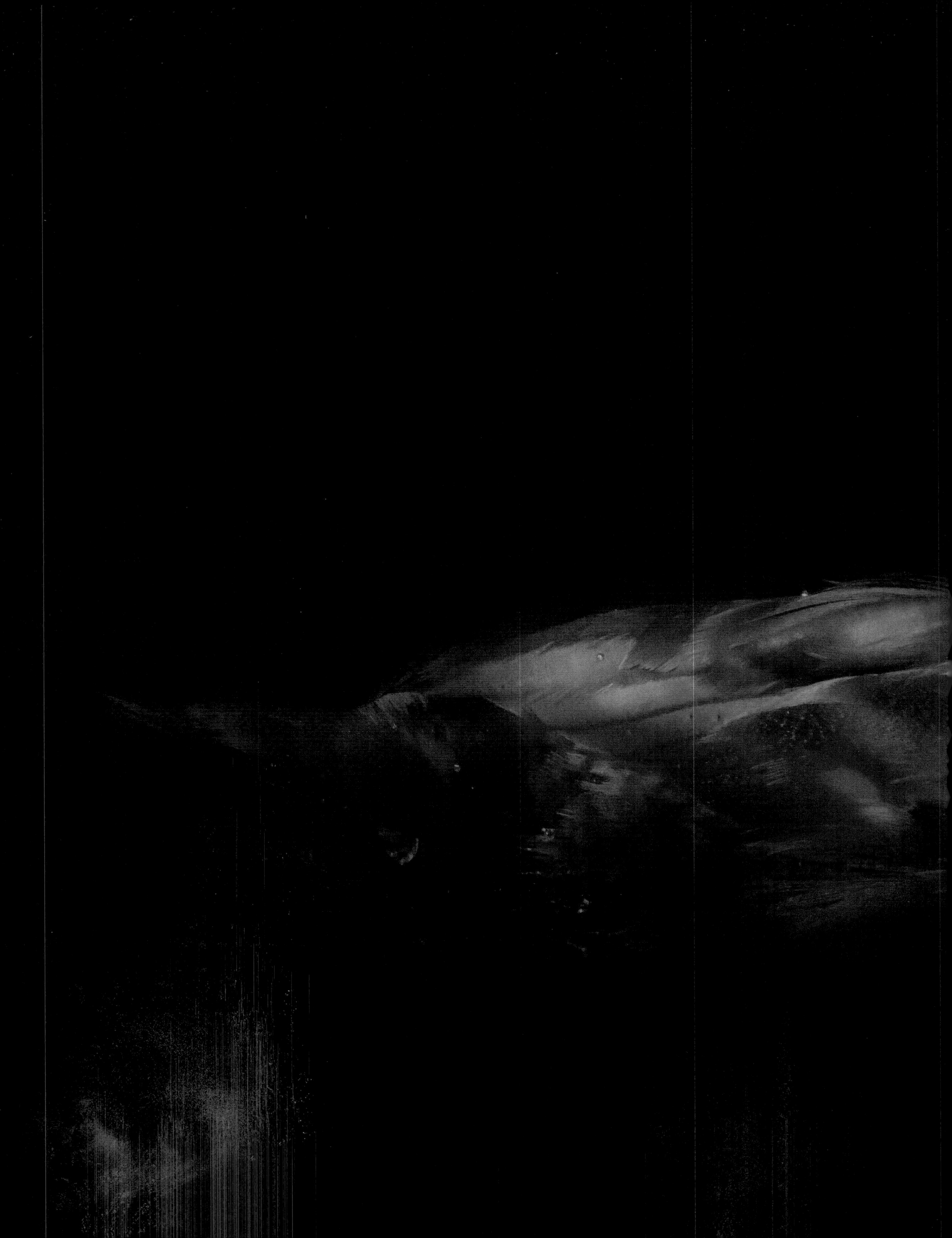

ABOVE
MOHO PERERŪ | BANDED RAIL
Gallirallus philippensis **| Declining**

OPPOSITE
SPUR-WINGED PLOVER
Vanellus miles novaehollandiae
Not threatened

Banded Rail live among tangled masses of grass, sedges, and rushes, which border swamps and lagoons. Owing to their shy and retiring disposition, they are seldom seen. They run through the vegetation with great agility and never take to flight except when on the point of being captured by a dog. Even then they fly a short distance only, with a slow flapping of the wings.

F. W. Hutton and J. Drummond, *The Animals of New Zealand: An Account of the Dominion's Air-breathing Vertebrates*, (Whitcombe and Tombs, 1909).

MATUKU-HŪREPO
AUSTRALASIAN BITTERN
Botaurus poiciloptilus **| Nationally critical**

To the Editor. Sir, — In the course of his paper read at the Philosophical Institute meeting last night, Mr Hutchinson said the bittern is fast becoming rare (thanks to the pot-hunter), and he had only once seen one actually on the Scinde Island side of the swamp channel. Surely the one he refers to can be seen at any time in an undertaker's window in Emerson street. Had Mr Hutchinson taken the trouble to look before he wrote the article he would have had a better idea of what he was talking about. As a shooter of long standing on these swamps I can assure him that at the beginning of the present shooting season there have been as many bitterns as ever there were. Had Mr Hutchinson gone still a little further with his research into the habits of the bittern he would have found that it was not the work of the pot-hunters, as he calls them. Sportsmen do not visit these swamps, I presume, but nature's instinct takes the birds elsewhere to find food.

I might also inform him that previous to the swan coming into the swamp they had plenty of eels for food, but when the silt came the eels shifted — silt and swamp eels do not agree — and freshwater carp were left in their place. Now that the swamp has little or no water half the year the carp have all perished, and so the bittern's principal food is not to be found.

'The Habits of the Bittern', *Daily Telegraph* (Napier), Issue 9811, July 10, 1900.

PŪKEKO
Porphyrio melanotus
Not threatened

When feeding, the pūkeko often lifts the food to its mouth with its claw, as a parrot does. It's flesh is very palatable, and the bird is, therefore, the object of many shooting expeditions. It feeds chiefly in the early morning and in the evening, hiding away among the flax bushes in the daytime. Vegetable substances form its principal food, but it also eats insects. The pūkeko is easily tamed and, when kept with domestic fowls, makes friends with them.

F. W. Hutton and J. Drummond, *The Animals of New Zealand: An Account of the Dominion's Air-breathing Vertebrates*, (Whitcombe and Tombs, 1909).

SOUTH ISLAND TAKAHĒ
Porphyrio hochstetteri
Nationally vulnerable

In 1849, some sealers who were pursuing their vocation at Duck Cove, Resolution Island saw the trail of a large and unknown bird on the snow, with which the ground was then covered.

They followed the footprints until they obtained a sight of the *notornis.* Their dogs at once pursued it, and, after a long chase, caught it alive in the gully of a sound behind Resolution Island. It ran with great speed and, on being captured, uttered loud screams and fought and struggled violently. It was kept alive for three days, on board the schooner, and was then killed. The precious body was roasted and eaten by the crew, each partaking of the dainty, which was declared to be delicious.

F. W. Hutton and J. Drummond, *The Animals of New Zealand: An Account of the Dominion's Air-breathing Vertebrates,* (Whitcombe and Tombs, 1909).

Estuary & Shore

'Certain persons have commenced to capture these seabirds and boil them down to extract oil from them.'

An interesting petition was presented, by Sir George Grey, from Sir James Hector and some twenty other well-known residents of Wellington, praying the House to pass an Act regulating the capture and destruction of seabirds on main and outlying islands of New Zealand. The petitioners particularly submit—(1) That the capture of seabirds for the purpose of extracting oil should be prohibited, and (2) That except for food, the capture of seabirds at the breeding places for their skins and feathers should be prohibited

The petition is a long one and gives a lot of curious information on the subject about which it treats. After detailing the habits of the penguins and other seabirds on the islands about New Zealand, the petition states that certain persons have commenced to capture these seabirds 'and boil them down to extract oil from them.' The most numerous species are the penguins on the Bounty Group, and of these the number is not estimated to exceed three million. The penguins in all the islands together are estimated not to exceed six million.

All these birds are gathered together in easily, accessible rookeries, and a party of four or five men could, the petitioners assert, with ease in one day capture every penguin on the Snares, in a few weeks all of the Auckland group, in two or three days all on Campbell Island, in one day all on the Antipodes, in a fortnight all on the Bounty Group, in a short additional time would virtually clear the seas of these birds, as they all come on shore to breed.

The amount of oil to be obtained by all this destruction would be insignificant, not more than 1000 tonnes.

The petitioners assert that certain persons visit island after island for the purpose of boiling down all the penguins thereon, and that if they are allowed to do so these birds will shortly be exterminated. The petitioners also believe that terns, gulls, albatrosses and particularly mutton birds, which are now a valuable article of food, will be treated in the same way. The petitioners also complain of eggs being taken from the nesting grounds of the albatross. Sir George Grey is moving to have the petition printed.

'A Commendable Movement: The Destruction of Seabirds', *Auckland Star*, Volume XXII, Issue 192, August 15, 1890.

TŌREA PANGO
VARIABLE OYSTERCATCHER
Haematopus unicolor | Recovering

TOP
TŪTURIWHATU
NORTHERN NEW ZEALAND DOTTEREL
Charadrius obscurus aquilonius
At risk / recovering

BOTTOM
POHOWERA
BANDED DOTTEREL
Charadrius bicinctus
Declining

OPPOSITE
POAKA | PIED STILT
Himantopus himantopus
Not threatened

ABOVE
KAWAU TIKITIKI | SPOTTED SHAG
Phalacrocorax punctatus | Nationally vulnerable

OPPOSITE
KĀRUHIRUHI | PIED SHAG
Phalacrocorax varius | Recovering

MĀPUNGA | BLACK SHAG

Phalacrocorax carbo | **Relict**

Mr C. J. Wilson has again devoted himself to the task of destroying the remainder of the colony of black shags, which have hitherto been a pest to the interests of acclimatisation in the Lake Ellesmere district. The day being hot, the birds flew low, and Mr Wilson was enabled during Thursday and the early part of Friday to kill no less than 105 black shags with his own gun. As they invariably fell into the water, he had no opportunity of opening them to learn their contents. This makes a total of 385 black shags which have been destroyed by Mr Wilson and his former comrade (Mr Lewis), who is certainly deserving of the thanks of the Colony for their exploitation. Finding the black shag had taken their departure, Mr Wilson turned his attention to the grey shag, which infests the sea coast but is not supposed to do any damage among freshwater fish. Of these, he succeeded during Friday afternoon in destroying 136, moving a total of 241 birds shot with one gun in two days. Mr Wilson considers that the Colony is so far exterminated that there will be no more black shags seen in the neighbourhood for some time to come, and the few survivors will now probably leave there.

'Black Shag Shooting,' *Star* (Christchurch), Issue 6710, November 25, 1889.

TĀKAPU | AUSTRALASIAN GANNET
Morus serrator **| Not threatened**

During the passage of the *Stella* from Wellington to Auckland, White Island was visited and was found to be in a state of great activity, throwing up stones, mud and steam. Some very handsome gannets were taken from the southern side of the island, and a large quantity of sea fowl's eggs of nearly every variety; these birds and fledglings and eggs are intended for the Auckland Museum.

'Interesting Cruise,' *Colonist*, Volume XXXII, Issue 5386, December 3, 1888.

MATUKU MOANA | WHITE-FACED HERON

Egretta novaehollandiae **| Not threatened**

I would fain write a whole book about the heron alone — percher, as well as swimmer and wader, is he. From what I have seen of him in the trees, he is as much at home in them as is a blackbird or a thrush, and his movements are far more graceful, notwithstanding his form and size; for the bird drops, glides and walks about the trees as noiselessly as a grey and white shadow. On the very topmost shoots of a fir that can afford him a foothold he will perch, and sit there for twenty minutes and more at a time. Recently I watched one do this from a place of ambush, which was, like most of those suitable for the purpose, anything but a pleasant one, being a large clump of blackthorns and brambles, in the midst of which I lay hidden. He was not, however, allowed to enjoy his elevated point of survey in peace, for missel-thrushes, blackbirds and song thrushes, aided by finches, dashed at him, until the long neck and bill went darting and striking out like a silvery snake in all directions as the birds flew at him. Then he was unmolested, and away they flew to cover, screaming and chirping their loudest.

'Herons', *Bruce Herald*, Volume XXIV, Issue 2495, July 21, 1893.

KŌTUKU | WHITE HERON

Ardea alba **| Nationally critical**

When the first settlers came to the South Island, this bird was not uncommon on the margins of lakes and rivers. Lake Heron, in the Ashburton County, was named by Mr Potts, in 1857, on account of the number of these birds which frequented its shores. But the white heron is now very rare in New Zealand. It has fallen prey to the gunner, who has killed it, not for food but in a spirit of wanton destructiveness.

F. W. Hutton and J. Drummond, *The Animals of New Zealand: An Account of the Dominion's Air-breathing Vertebrates*, (Whitcombe and Tombs, 1909).

ABOVE
TARĀPUNGA RED-BILLED GULL (JUVENILE)
Chroicocephalus novaehollandiae scopulinus | Declining

OPPOSITE
KARORO BLACK-BACKED GULL
Larus dominicanus | Not threatened

To the Editor. Sir, — One of the prettiest ornaments of our harbour is the flights of seabirds which frequent it. The little seagulls especially, finding food near the dwellings of men, and by comparative immunity rendered tame, flock all around from Te Aro shore to Kaiwharawhara at low water, like flakes of their own sea foam. I am sorry to say that lately, almost every day, barbarians on shore and in boats are massacring these innocent, beautiful little creatures from sheer wanton love of destruction; for they are of no use for eating. Surely the killing of these seabirds in the inner harbour is against the law; and if so, the penalty should be sternly exacted, and that before they are frightened away forever. And the same thing should be done with respect to the tūī, the only bird which relieves the New Zealand bush from painful impressions of silence and solitude. I have lately seen some Cockney sportsmen come into a farmhouse with a whole bagful of murdered tūī, mere bloodstained bunches of feathers. I think that in these cases anyone who witnesses the killing of seagulls about town, or the tūī in the country, should inform against the perpetrators of a cruel outrage on the laws of humanity.

'Killing Seagulls,' *New Zealand Mail*, Issue 271, April 7, 1877.

RIGHT
TARA
WHITE-FRONTED TERN
Sterna striata | Declining

OVERLEAF
TARAPIROHE
BLACK-FRONTED TERN
Chlidonias albostriatus
Nationally endangered

The beautiful little bird was once seen in large numbers hovering over the newly-ploughed fields in search of larvae of various insects and small lizards, but it is rarely seen now. The cause of the decline in numbers is a mystery.

F. W. Hutton and J. Drummond, *The Animals of New Zealand: An Account of the Dominion's Air-breathing Vertebrates*, (Whitcombe and Tombs, 1909).

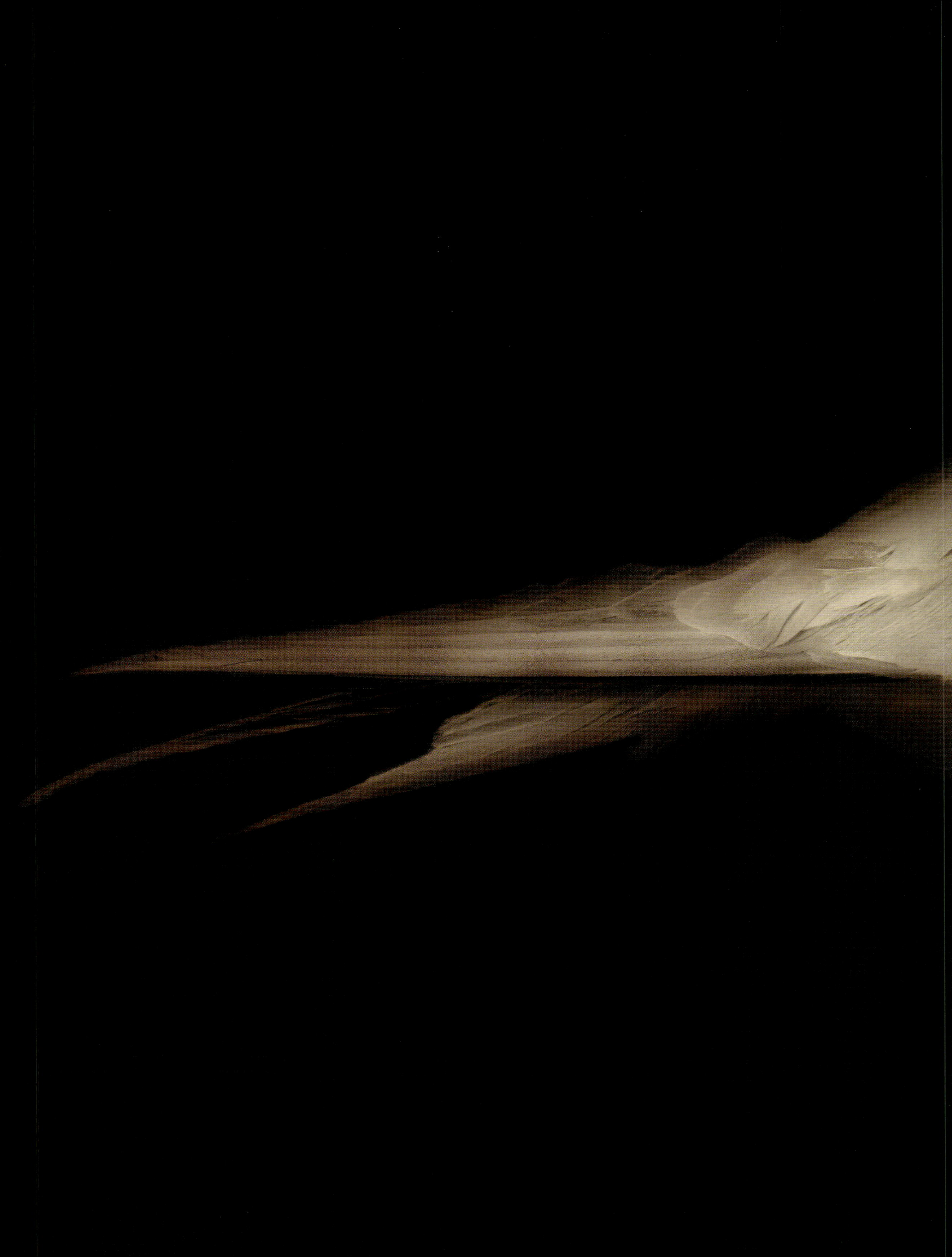

ABOVE LEFT

KORORĀ | LITTLE PENGUIN

Eudyptula minor **| Declining**

Sir, — In the account of the Auckland, Campbell and other Islands, which appears in your issue of September 9, your special correspondent writes as if penguins were unknown on the mainland of New Zealand. This is not correct. Since 1874 I always found them numerous in the sounds and along the coast between Martin's and Jackson's Bays, and Captain Cook mentions seeing penguins when he sailed around the mainland in 1770. On the West Coast, I have often come on them in the bush some distance from the sea. I should say there are as many penguins on the West Coast as there are on the islands mentioned, only having more room on the coast they do not crowd together like they do on the islands. The disappearance of the penguins mentioned by your correspondent is probably due to the action of some gentlemen who have recently been trying to make a trade of boiling down penguins for their oil.

Robert Paulin, 'Penguins: To the Editor,' *Otago Witness*, Issue 2017, October 20, 1892.

OPPOSITE

HOIHO | YELLOW-EYED PENGUIN

Megadyptes antipodes **| Nationally endangered**

ABOVE RIGHT

TAWAKI | FIORDLAND CRESTED PENGUIN

Eudyptes pachyrhynchus **| Declining**

TOROA
SOUTHERN ROYAL ALBATROSS
Diomedea epomophora
Nationally vulnerable

That the fauna of your lovely country is exceedingly rich is acknowledged, yet no one had any idea that so many useful articles could be manufactured from the feathers of your birds or the skins of your small animals. I have already in a former letter alluded to H. E. Liardet's collection of stuffed birds and animals. His case of goods made from them — says my lady informant and assistant (for in such matters what man could be trusted?) — are of highly satisfactory quality and finish. A muff and collar of pure white from the albatross is fit (she says) for 'the first lady in the land.' Muffs and collarettes are here made of the spotted shag, exceedingly pretty. Others made out of the paradise duck would make ladies hope that they may be in fashion when they enter the promised abode of bliss. A charming specimen, made out of the feathers of the shoveler duck, is of a dusky brown colour.

'New Zealand Furs', *Auckland Star*, October 21, 1879.

We must not forget to mention the pretty tobacco pouches made out of the web foot of seabirds, which afford a capital opportunity to reciprocate such presents of plumage as they may receive. The albatross pouch, indeed, carries a lady's handkerchief well.

North Otago Times, Volume XX, Issue 952, July 10, 1874.

Extinct

It is the opinion of the highest authorities that at no distant period, it will be impossible to procure a collection of many species, even of the common birds now found in this country; before long, these too will have disappeared with the moa.

'His Excellency Sir G. F. Bowen at the Formal Opening of the New Zealand Institute,' *Colonist*, Volume XI, Issue 1142, September 4, 1868.

In the early days, our forests were peopled by thousands of birds of beautiful plumage and peculiar characteristics, and now a person may roam through the bush for weeks and not see more than a few tūī, parakeets, linnets and perhaps a stray robin, woodhen or fantail.

What has become of all the other useful and ornamental birds, and why are the few kinds occasionally seen dying out so rapidly? The disciples of the association that imported grain-eating birds at a cost of hundreds of pounds, half of which (if I am not misinformed) was paid out of the public treasury, may answer: The native birds have been slain by thousands, are being put to death by hundreds, and will soon be annihilated by the people of the country through their guns, bush-clearing and the dogs, cats and vermin they introduced.

Fitz Barry, 'The Acclimatisation Society and Native Birds,' *Otago Witness*, Issue 1533, March 26, 1881.

KAORIKI | NEW ZEALAND LITTLE BITTERN

Ixobrychus minutus novaezelandiae **| Extinct 1890s**

It is stated that the little bittern is so quiet in its habits that it will remain still when approached so that it will apparently ensure itself to be taken by the hand. The birds just described were taken alive, it has been said, without any very great difficulty; after which they were turned loose among the fowls in a poultry yard. They were found dead shortly afterwards, it is alleged, from exposure to the keen frosty night air, being deprived of the accustomed protection afforded by the thickly-growing sedgy vegetation of their swampy habitat.

I have seen them in two positions, viz., standing on the bank of the lagoon, with their heads bent forward, studiously watching the water; at other times I have seen them standing straight up, almost perpendicular — I should say this is the proper position for the bird to be placed in when stuffed.

Walter L. Buller, 'Further Notes on the Birds of New Zealand', *Transactions and Proceedings of the Royal Society of New Zealand* 3, (1870): 59.

HUIA

***Heteralocha acutirostris* | Extinct 1900s**

The huia are among the remarkable birds that make the fauna of New Zealand especially interesting. Their most striking peculiarity to a naturalist lies in the difference between the bills of the male and the female. The male's bill is straight, while the female's is strikingly curved and has a considerable advantage in regard to length. After an extensive journey in the hilly forest in search of them, I had, at last, the pleasure of seeing four alight on the lower branches of the trees near which the guide accompanying me stood. They came as quick as lightning, descending from branch to branch, spreading out their tails, and throwing out their wings. Anxious to obtain them, I fired; but they generally come so near that hunters can kill them with sticks.

F. W. Hutton and J. Drummond, *The Animals of New Zealand: An Account of the Dominion's Air-breathing Vertebrates*, (Whitcombe and Tombs, 1909).

To show how much scarcer this bird is than it was formerly, I may mention that a few months ago, accompanied by Mr Morgan Carkeek, I made an expedition into the wooded ranges at the back of Waikanae. We crossed the Akatarawa saddle into the valley of the Hutt, and made a ten-mile circuit over the wooded ranges, cutting our path with bill-hooks through the virgin forest, rendered almost impervious by a tangle of kiekie and supplejack, and camped several nights in the woods. During the whole expedition we only saw a single huia — a male bird, which visited our camp in the early morning. Mr Carkeek assures me that when exploring and surveying in these ranges only five or six years ago the huia was comparatively plentiful.

On a more recent occasion, accompanied by Māori, I visited the portion of the Forty-mile Bush where, as related in my *Birds of New Zealand*, I obtained so many specimens in 1883. But the bush has completely disappeared before the advancing wave of European settlement.

From Pahiatua we rode for twenty miles through clearings exhibiting nothing but charred and naked stumps, the whole of this country being at the time of my former visit covered with beautiful forest. From the practical standpoint of material advancement there is nothing regrettable in this; but the fact remains that the home of the huia is being swept away, and although these birds in greatly-diminished numbers, have taken refuge in the wooded mountain ranges, the date of their extinction cannot be very far distant.

Walter L. Buller, 'Further Notes on the Birds of New Zealand,' *Transactions and Proceedings of The Royal Society of New Zealand* 25, (1892): 63.

WHĒKAU | LAUGHING OWL

Ninox albifacies **| Extinct 1910s**

This fine owl is now on the verge of extinction; indeed, I have made such persistent efforts all over the country to obtain specimens, since my return from Europe about three years ago, that I think it highly probable the single live pair which I have been fortunate enough to procure will be the last we shall ever get. These have been sent to England and will be kept in an aviary specially prepared for them. Seeing that the single bird belonging to the Canterbury Acclimatisation Society lived in confinement for some eighteen years, and laid an unfertilised egg (now in the Canterbury Museum) shortly before its death, we may reasonably hope that my captive birds will breed in their new home, and that in this way the race will be perpetuated.

Mr Jacobs, the taxidermist at Masterton, has given me a characteristic account of a very fine one which he obtained alive at Nelson, and afterwards mounted for the local Museum. A man, so he informed me, was travelling from Nelson to the West Coast, when he observed a large owl squatting on the ground near the roadside. He dismounted from his horse and caught the bird. Then, selecting a retired nook in the adjoining woods, he drove a thick pole into the ground and secured his captive to it by the leg, allowing a sufficient length of flax to permit the owl to move freely about over the ground. On his return by the same road two days later he found that the bird had snapped, or in some way had got disengaged from, the flax string, and was perched on the top of the pole, permitting itself to be recaptured without the slightest resistance. He took it on with him to Nelson, and, not knowing its value, sold it to the narrator for a few shillings. It now graces the collection in the Nelson Museum.

Walter L. Buller, 'Further Notes on the Birds of New Zealand,' *Transactions and Proceedings of the Royal Society of New Zealand* 25 (1892):53.

SOUTH ISLAND PIOPIO

Turnagra capensis **| Extinct 1900s**

The eye of the piopio seems to gleam with intelligence. Both skin and flesh of the bird are dark, but it is stated that the flavour of the flesh is not at all unpleasant.

'It makes a savoury broil for those who bring the proper sauce — hunger,' adds Mr Potts.

F. W. Hutton and J. Drummond, *The Animals of New Zealand: An Account of the Dominion's Air-breathing Vertebrates*, (Whitcombe and Tombs, 1909).

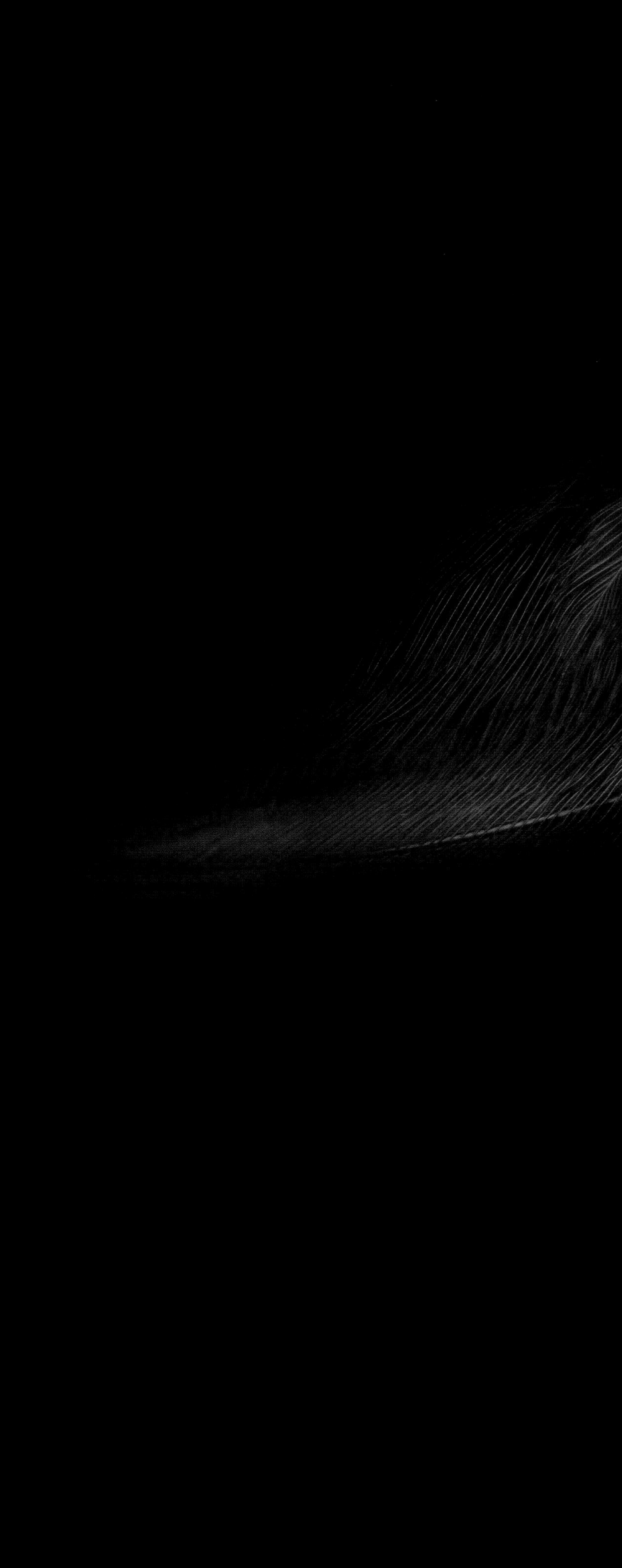

MĀTUHITUHI | BUSH WREN
Xenicus longipes **| Extinct 1970s**

Mr Brough writes me from Nelson, 'I have now been out in the bush for six months, and have seen only one bush wren, two rifle wrens, two saddlebacks and no rock wrens. These birds are almost extinct in the Nelson and Pelorus forests, where they were so plentiful eight years ago. Weasels, ferrets, stoats, rats and wild cats abound in our woods. Cats are the greatest enemy to the wren family. The animals I have mentioned are making terrible devastation among our native birds. Wingless birds, and pigeons too, will soon be a rarity here.'

Walter L. Buller, 'Further Notes on the Birds of New Zealand,' *Transactions and Proceedings of the Royal Society of New Zealand* 25, (1892): 63.

KŌKĀ | SOUTH ISLAND KŌKAKO

Callaeas cinerea **| Extinct 2000s (Declared)**

Dr. Cahill kindly presented me with a live bird which he had received from Westport, and I had it in my possession many months. Finally it died in a fit, due, I think, to overfeeding without the stimulus of freer exercise.

It was an adult male and in perfect plumage, with bright orange wattles, dark blue at the base. Its habitual note, emitted frequently, but chiefly in the early morning and forenoon, was a long, plaintive double-note, pitched in a minor key, very pleasant to hear, but to my mind possessing less richness than the organ-note of the North Island bird (*G. wilsoni*). It was accustomed to use its feet on eating leaves or berries presented to it, just as a parrot would.

On offering this bird a large blue-bottle fly he held it to his perch in the manner described, and deliberately tore off one wing, then the other, tasted its flavour, and immediately dropped it. As a rule he would not touch insects, but showed great fondness for succulent leaves of any kind, particularly those of *Coprosma lucida*, and all sorts of native berries, whether ripe or green.

He ate freely of the ripe fruit of the whauhe, but it had a scouring effect, and I had to discontinue the use of this food. It partook readily of cooked potato, boiled rice and soaked bread; and it was fond of water, drinking freely, but rarely washing itself as other birds do, and yet its plumage was always in clean, silky condition.

Walter L. Buller, Further Notes on the Birds of New Zealand,' *Transactions and Proceedings of The Royal Society of New Zealand* 25, (1892): 63.

Chronology of Extinctions

Moeriki Dieffenbach's Rail

1840s

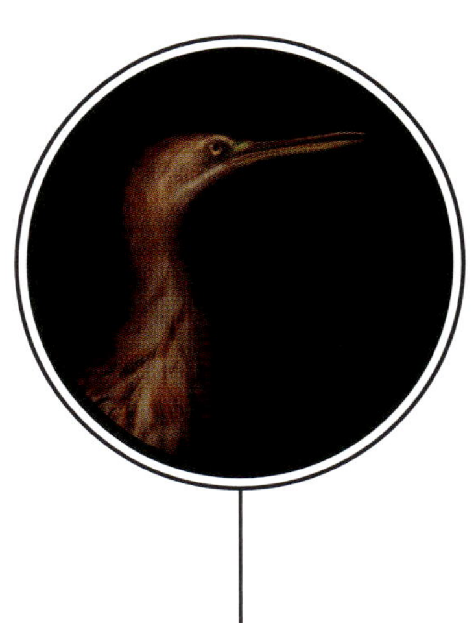

Chatham Island Fernbird
Kaoriki New Zealand Little Bittern
Lyall's Wren
Mātirakahu Chatham Island Rail

1890s

1870s

Koreke New Zealand Quail
North Island Snipe

1900s

Huia
Kōmako Chatham Island Bellbird
Miuweka Auckland Island Merganser
North Island Piopio
South Island Piopio

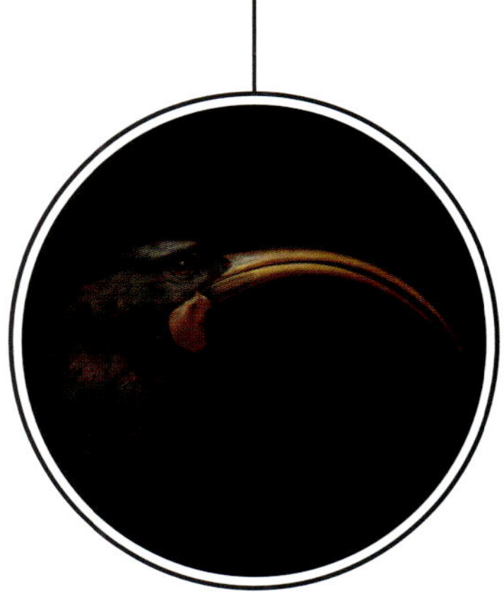

Whēkau Laughing Owl

1910s

Kōkā South Island Kōkako
(declared)

2000s

1960s–1970s

Mātuhituhi Bush Wren
Tutukiwi South Island Snipe

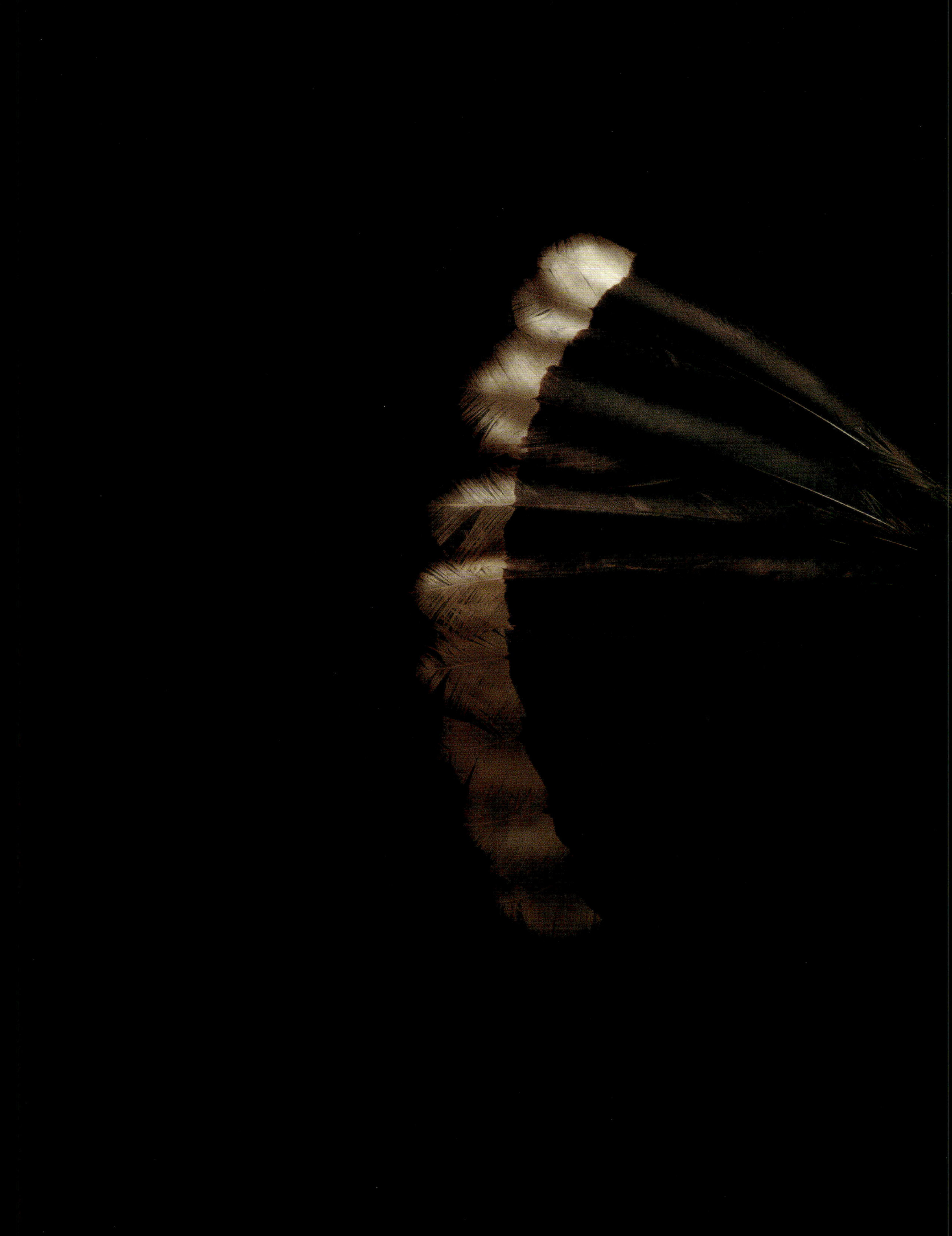

Acknowledgements

This book is the culmination of decades of work. I would like to express my deepest gratitude to everyone who has contributed in some way to its creation.

First and foremost, I want to thank my family, especially my husband, Greg, whose unwavering support and encouragement (and the countless kilometres he has driven, hiked and rowed with me to reach my subjects) have made this project possible. To my children, who grew up alongside this book, Ethan, who first sparked the idea in me, and Willow, my wonderful companion on so many photo shoots from a young age – thank you for your patience and understanding through countless cries of 'Just one more shot.' Lastly, to my father, who did not live to see this book completed but who, from my earliest years, instilled in me a love of nature and an eye for light and shadow.

Special thanks to my editor, Jemma Moreira, whose thoughtful feedback and meticulous attention to detail were invaluable as we crossed the finish line, and to designer Megan van Staden, who patiently and faithfully created the look I was after. I also extend my gratitude to the team at Bateman Books, including Paul Bateman and Louise Russell, for believing in this project and bringing it to life, and to Nicola Toki for taking the time to write a foreword to support this project.

I am deeply grateful to Te Papa Tongarewa, Auckland War Memorial Museum Tāmaki Paenga Hira, and the National Library of New Zealand Te Puna Mātauranga o Aotearoa for generous access to their archives and collections, which were critical to the research and historical extracts within these pages. Thanks also to The Animal Sanctuary in Matakana for their kindness in allowing me access to their native birds.

My heartfelt thanks go to Richard Loseby for encouraging me to embark on this publishing journey, and to friends and family for their constant encouragement and faith in this project.

I have photographed countless ducks, pigs, sheep, cows, goats and chickens over the years – a warm thank you to all my neighbours who so generously allowed me onto their land to do so.

Finally, I want to extend my deepest gratitude to all those who work tirelessly towards a predator-free Aotearoa New Zealand. Your dedication to undoing the damage of the past is a beacon of hope for our native wildlife. Without your efforts, the future of our birdlife would be all but sealed.

About the Artist

Cally Whitham is a photographer based in Aotearoa whose work seeks to preserve the overlooked and the ordinary by transforming them into something timeless. Her evocative images capture the subtle beauty of introduced and native birds, farm animals and rural scenes, often bathed in natural light that evokes a sense of memory and nostalgia.

Cally's journey with photography began at the age of eleven, using her first roll of film to document cherished surroundings — her aunt's farm, old houses and trees. This early fascination with capturing the fleeting moments of life has stayed with her, evolving into a body of work that reflects both her personal story and the shifting landscapes of Aotearoa.

Her lineage traces back to some of the country's earliest settlers in 1834; a history that deeply informs her exploration of acclimatisation, heritage and the complex interplay between people and nature. These themes resonate in her art, which has reached a global audience. Her work has become widely popular online, leading to interviews with countless blogs and magazines including features in *National Geographic*, *GEO magazine*, *Internazionale* and more.

Living on a lifestyle block north of Auckland surrounded by the subjects that inspire her, Cally continues to create images that reflect the enduring relationship between humans and the natural world. Her photographs invite viewers to pause and consider the beauty in what we often take for granted, offering a poignant reminder of the stories held within familiar landscapes.

Bibliography

BOOKS

Charles Hursthouse, *New Zealand, or, Zealandia, the Britain of the South, Vol. 1*, (Edward Stanford, 1857).

F. W. Hutton and J. Drummond, *The Animals of New Zealand: An Account of the Dominion's Air-Breathing Vertebrates*, (Whitcombe and Tombs, 1909).

G. M. Thomson, *The Naturalisation of Animals and Plants in New Zealand*, (Cambridge University Press, 1922).

G. M. Thomson, *Wild Life in New Zealand, Part 1, Mammalia*, (Wellington Government Printer, 1921).

J. Drummond, *Our Feathered Immigrants: Evidence for and Against Introduced Birds in New Zealand, Together with Notes on the Native Avifauna* (Wellington Government Printer, 1907).

JOURNALS

From *Papers Past*, National Library of New Zealand Te Puna Mātauranga o Aotearoa:

'Inaugural Address of Governor Sir George Ferguson Bowen', *Transactions and Proceedings of the Royal Society of New Zealand* 1, (1868):1.

J. Drummond, 'On Introduced Birds', *Transactions and Proceedings of the Royal Society of New Zealand* 39, (1906): 227.

'On the Birds of New Zealand', *Transactions and Proceedings of the Royal Society of New Zealand* 3, (1870): 59.

Walter L. Buller, 'Further Notes on the Birds of New Zealand', *Transactions and Proceedings of the Royal Society of New Zealand* 3, (1870): 59.

Walter L. Buller, 'Further Notes on the Birds of New Zealand', *Transactions and Proceedings of the Royal Society of New Zealand* 25, (1892): 63.

ARTICLES

From *Papers Past*, National Library of New Zealand Te Puna Mātauranga o Aotearoa:

'Acclimatisation', *Daily Southern Cross,* Volume XXIV, Issue 3343, April 3, 1868, 3.

'Acclimatisation', *Otago Witness*, May 26, 1866.

'A Commendable Movement: The Destruction of Seabirds', *Auckland Star*, Volume XXII, Issue 192, August 15, 1890, 3.

Agricola, 'The Morepork', *Daily Southern Cross*, Volume XXIV, Issue 3382, May 19, 1868, 3.

A. Gunner, 'The Naturalist: Some Notes on Acclimatisation', Gardner, *Otago Witness*, Issue 2013, June 14, 1894, 49.

'A Hedgeho' [sic], *Wanganui Chronicle*, Volume XXV, Issue 10410, September 24, 1883.

Alex C. Begg, 'Councillor Robin and the Acclimatisation Society's Ducks', *Otago Daily Times*, Issue 7317, July 29, 1885.

'A Trip to the Tableland: Range of Birds and Taxidermy', *Nelson Evening Mail,* Volume XXXII, Issue 208, September 9, 1898, 3.

Auckland Star, Volume VI, Issue 1777, October 25, 1875, 2.

Auckland Star, Volume XXIV, Issue 222, September 20, 1898.

'Black Shag Shooting', *Star* (Christchurch), Issue 6710, November 25, 1889, 2.

'Bush Dogs in the North', *Taranaki Herald*, Volume XV, Issue 748, December 1, 1866, 2.

'Carrier-Pigeon Post: A System Known to the Ancients', *Mataura Ensign*, Issue 871, March 28, 1901, 4.

C. T. Wren, 'The Arabian Millet', *New Zealand Herald*, Volume XIX, Issue 6563, November 29, 1882.

C. W. Purnell, 'The Takahē', *Ashburton Guardian*, January 24, 1889, 2.

'Destruction of Hawks', *Lyttelton Times*, Volume LXI, Issue 7240, May 14, 1884.

'Farmers' Column', *Patea Mail*, November 10, 1882, 2.

Fitz Barry, 'The Acclimatisation Society and Native Birds', *Otago Witness*, Issue 1533, March 26, 1881, 12.

'Furious Riding', *Southland Times*, Issue 12424, October 14, 1892, 3.

H. A. Bruce, 'Protected Birds', *Lyttelton Times,* Volume XCIII, Issue 10616, March 27, 1895, 3.

'Hares v. Rabbits', *Wairarapa Standard*, Volume XVII, Issue 1443, August 15, 1883, 3.

Hawke's Bay Weekly Times, Volume 2, Issue 70, May 4, 1868, 108.

H. B. Huddleston, 'The Caterpillar Plague: To the Editor', *Bay of Plenty Times*, March 20, 1899, 2.

'Herons', *Bruce Herald*, Volume XXIV, Issue 2495, July 21, 1893, 5.

'His Excellency Sir G. F. Bowen at the Formal Opening of the New Zealand Institute,' *Colonist*, Volume XI, Issue 1142, September 4, 1868.

'How the Premier Lost His Jackass', *The Press,* Volume LVIII, Issue 11020, July 18, 1901, 5.

'Interesting Cruise', *Colonist,* Volume XXXII, Issue 5386, December 3, 1888, 4.

'Interprovincial', *Ashburton Guardian,* November 22, 1890, 2.

J. McElwain, 'Destruction of Birds in Waitākere Reserve', *Auckland Star,* Volume XXVI, Issue 107, May 6, 1895, 4.

'Killing Seagulls', *New Zealand Mail,* Issue 271, April 7, 1877, 20.

'Local and General', *Ashburton Guardian*, January 24, 1889.

Mark P. Stoddart, 'Acclimatisation', *Lyttelton Times*, February 18, 1864, 5.

'Miscellaneous', *New Zealand Mail,* September 8, 1898, 31.

'New Zealand Furs,' *Auckland Star,* Volume X, Issue 2970, October 21, 1879, 2.

New Zealand Mail, Issue 988, February 6, 1891.

North Otago Times. Volume XX, Issue 952, July 10, 1874, 2.

'Notes from the Bush', *Wellington Independent*, Volume XIX, Issue 2166, February 11, 1865, 10.

Oamaru Bruce Herald, November 21, 1873.

'Opossums in New Zealand', *The Press,* January 25, 1896, 7.

'Our Native Game', *Otago Witness*, Issue 2455, April 3, 1901, 48.

Patea Mail, Volume VIII, Issue 957, November 10, 1882.

'Protection of Game', *New Zealand Mail,* Issue 481, April 30, 1881, 14.

'Provincial and General', *North Otago Times*, Volume XIII, Issue 490, February 4, 1870, 3.

R. Henry, 'Exportation from Dusky Sound', *Otago Daily Times,* Issue 10613, March 7, 1896, 6.

Robert Paulin, 'Penguins: To the Editor', *Otago Witness*, Issue 2017, October 20, 1892, 21.

'The Habits of the Bittern', *Daily Telegraph* (Napier), Issue 9811, July 10, 1900, 5.

'The Paradise Duck', *Otago Witness,* Issue 1921, December 11, 1890, 32.

'The Story of Our Plants and Birds', *The Press*, Volume LVII, Issue 10840, December 15, 1900, 20.

'The Takahē', Purnell, C. W., *Ashburton Guardian*, January 12, 1889.

Waikato, 'Pheasant: Its Destructive Habits', *Daily Southern Cross,* January 15, 1868, 3.

Walter L. Buller, 'The Vanishing Forms of Bird-Life in New Zealand', *The Press*, Volume LIV, Issue 9622, January 11, 1897.

'The Wingless Birds of New Zealand', *New Zealander,* Volume 7, Issue 550, July 23, 1851, 4.

'Town & Country', *Lyttelton Times*, Volume LXXV, Issue 9331, February 6, 1891.

'Town & Country', *Timaru Herald*, Volume LXXIX, Issue 12190, October 7, 1903, 2.

Lyttelton Times, Volume XXI, Issue 1198, February 18, 1864.

Otago Witness, Issue 756, May 26, 1866.

Wellington Independent, Volume XIX, Issue 2166, February 11, 1865, 10.

Image Credits

I have created portraits from taxidermy specimens in cases where live birds or animals were not accessible for photography. Many of these specimens were collected during early European settlement, intended for museums and cabinets of curiosity. Their inclusion not only enhances the authenticity of this book but also reflects the practices and attitudes of the time, providing a historically relevant and visually compelling narrative.

Special thanks to Museum of New Zealand Te Papa Tongarewa and Auckland War Memorial Museum Tāmaki Paenga Hira for making their collections available for public use. Additional thanks to the Forest Bridge Trust for access to their taxidermy specimens.

Below is a list of images obtained under a 4.0 Creative Commons license or sourced from the public domain. All works have been significantly altered from their original form. No AI was used in the creation of images for this book.

MUSEUM OF NEW ZEALAND TE PAPA TONGAREWA

Pages iv–v & 130–131: North Island Brown Kiwi *Apteryx mantelli*, collected Northland, New Zealand. Gift of New Zealand Wildlife Service, –1987. CC BY 4.0. Te Papa (OR.023841)

Pages vi & 122–123: Kākāpō *Strigops habroptilus*, collected June 1992, Codfish Island, Foveaux Strait, New Zealand. Gift of the Department of Conservation, 1987. CC BY 4.0. Te Papa (OR.025736)

Page 7: Goldfinch *Carduelis carduelis britannica*, collected October 1979, Featherston, New Zealand. CC BY 4.0. Te Papa (OR.022042)

Pages 20–21: Little Owls *Athene noctua*, collected October 2, 1932, Oamaru, New Zealand. Gift of F. Milner, 1932. CC BY 4.0. Te Papa (OR.013971)

Page 25: Grey Partridge *Perdix perdix*, collected South Island, New Zealand. Gift of New Zealand Wildlife Service, 1977. CC BY 4.0. Te Papa (OR.019487)

Page 26: Brown Quail *Synoicus ypsilophorus australis*, collected no data. Gift of Mount Bruce Wildlife Reserve, 1999. CC BY 4.0. Te Papa (OR.025908)

Page 134: South Island Robin *Petroica australis*, collected Owen River Junction, Buller Valley, Nelson, New Zealand. Acquisition history unknown. Found in collection, date unknown. CC BY 4.0. Te Papa (OR.014054)

Page 134: North Island Rifleman *Acanthisitta chloris granti*, collected August 15, 1927, Tararua Mountains, New Zealand. Field Collection 1927. CC BY 4.0. Te Papa (OR.001916)

Page 135: Whitehead *Mohoua albicilla*, collected January 18, 1919, Ātene Pā, Whanganui, New Zealand. Purchased 1939. CC BY 4.0. Te Papa (OR.001724)

Page 145: North Island Kōkako Callaeas wilsoni, collected September 19, 1979, Mapara, New Zealand. Gift of New Zealand Wildlife Service, 1987. CC BY 4.0. Te Papa (OR.022072)

Pages 150–151: Buff Weka *Gallirallus australis hectori* (Hutton, 1873), collected Mount Bruce Wildlife Reserve, Wairarapa, New Zealand. Gift of New Zealand Wildlife Service, 1987. CC BY 4.0. Te Papa (OR.022234)

Pages 156–157: Kākāpō *Strigops habroptilus*, collected July 1993, Mana Island, Wellington, New Zealand. Gift of New Zealand Wildlife Service, –1987. CC BY 4.0. Te Papa (OR.025019)

Page 160: New Zealand Falcon *Falco novaeseelandiae*, collected June 6, 1937, Rotorua, New Zealand. Gift of A. Kean, 1937. CC BY 4.0. Te Papa (OR.013975)

Page 161: Australasian Harrier *Circus approximans*, collected June 1940, Wairarapa, New Zealand. Acquisition history unknown. Found in collection, date unknown. CC BY 4.0. Te Papa (OR.010596)

Page 164: New Zealand Dabchick *Poliocephalus rufopectus*, collected May 17, 1943, Pōrangahau, Eastern Hawke's Bay, New Zealand. Gift of J. L. Hunter, 1943. CC BY 4.0. Te Papa (OR.014141)

Page 174: Banded Rail *Gallirallus philippensis assimilis*, collected Marineland, Napier, New Zealand. Gift of Mount Bruce Wildlife Reserve, 1999. CC BY 4.0. Te Papa (OR.025910)

Page 186: Spotted Shag *Phalocrocorax punctatus*, collected 1912, Lyttelton Harbour, New Zealand. CC BY 4.0. Te Papa (OR.000827)

Pages 194–195: White Heron *Egretta alba modesta*, collected Wellington Zoo, Wellington, New Zealand. Gift of Wellington Zoo, date unknown. CC BY 4.0. Te Papa (OR.025204).

Pages 202–203: Black-fronted Tern *Chlidonias albostriatus*, collected May 22, 1951, South Wairarapa, New Zealand. Gift of A. C. O'Connor, 1951. CC BY 4.0. Te Papa (OR.000737)

Page 204: Yellow-eyed Penguin *Megadyptes antipodes*, collected no data, New Zealand. Acquisition details/history unknown. Found in collection, date unknown. CC BY 4.0. Te Papa (OR.014076)

Page 205: Fiordland Crested Penguin *Eudyptes pachyrhynchus*, collected no data, New Zealand. Gift of the Department of Conservation, 1987. CC BY 4.0. Te Papa (OR.024031)

Pages 206–207: Southern Royal Albatross *Diomedea epomophora*, collected no data. Acquisition details/history unknown. Found in collection, date unknown. CC BY 4.0. Te Papa (OR.014090)

Page 210: Huia *Heteralocha acutirostris*, collected no data, New Zealand. Purchased 1916. CC BY 4.0. Te Papa (OR.000152)

Pages 215 & 224: New Zealand Little Bittern *Ixobrychus novaezelandiae*, collected Head of Lake Wakatipu, New Zealand. Purchased 1907. CC BY 4.0. Te Papa (OR.000644)

Pages 216–217 & 225: Whēkau Laughing owl *Ninox albifacies*, collected no data, New Zealand. Purchased 2015. CC BY 4.0. Te Papa (OR.030061)

Pages 218–219: South Island Piopio *Turnagra capensis*, collected no data, New Zealand. Purchased 1912. CC BY 4.0. Te Papa (OR.000208)

Pages 220–221 & 225: South Island Bush Wren *Xenicus longipes longipes*, collected no data, New Zealand. Acquisition details/history unknown. Found in collection, date unknown. CC BY 4.0. Te Papa (OR.001334)

Pages 222–223 & 225: South Island Kōkako *Callaeas cinerea*, collected no data, New Zealand. Gift of the Wellington City Council, 1929. CC BY 4.0. Te Papa (OR.014041)

Pages 224 & 226–227: Huia *Heteralocha acutirostris*, collected no data, New Zealand. Acquisition details/history unknown. Found in collection, date unknown. CC BY 4.0. Te Papa (OR.000064)

AUCKLAND WAR MEMORIAL MUSUEM TĀMAKI PAENGA HIRA

Page 19: Rook *Corvus frugilegus*; LB8845; © Auckland Museum CC BY. This work is licensed under a Creative Commons Attribution 4.0 International License.

Page 25: Partridge *Chukor Alectoris chukar*, LB8835, © Auckland Museum CC BY. This work is licensed under a Creative Commons Attribution 4.0 International License.

Page 153: Parakeet Red Crowned *Cyanoramphus novaezelandiae*, LB4175, photographed by Denise Baynham, digital, January 17, 2017, © Auckland Museum CC BY. This work is licensed under a Creative Commons Attribution 4.0 International License.

Page 165: Australasian Grebe *Podiceps cristatus*, LB3978; © Auckland Museum CC BY. This work is licensed under a Creative Commons Attribution 4.0 International License.

Pages 176–177: Australasian Bittern *Botaurus poiciloptilus*, LB8847; © Auckland Museum CC BY. This work is licensed under a Creative Commons Attribution 4.0 International License.

Page 224: Koreke New Zealand *Quail Coturnix novaezelandiae*; LB4106; © Auckland Museum CC BY. This work is licensed under a Creative Commons Attribution 4.0 International License.

Cover & pages 214–215: Huia *Heteralocha acutirostris*, LB8536, © Auckland Museum CC BY. This work is licensed under a Creative Commons Attribution 4.0 International License.

PUBLIC DOMAIN

All images have been significantly altered from the original photo.

Page 4: Dunnock, pcjvdwiel, Pixabay
Page 4: Red Poll, Patrice Bouchard, Unsplash
Page 6: Chaffinch, David George, Unsplash
Page 11: Common Starling, Theotherkev, Pixabay
Pages 14–15: Rosella, David Clode, Unsplash
Page 29: Guineafowl, Sharkolot, Pixabay
Page 86: Wapiti, StockSnap, Pixabay
Pages 88 & 119: Chamois, Nina, Pixabay
Page 95: Hare, Angela, Pixabay
Page 188: Black Shag, Nick Sokolov, Unsplash
Page 189: Gannet, Elliott, Pixabay
Pages 210 & 224: Diffenbach's Rail, Wikimedia Commons

Published in 2025 by David Bateman Ltd,
Unit 2/5 Workspace Drive, Hobsonville,
Auckland 0618, New Zealand
www.batemanbooks.co.nz
ISBN: 978-1-77689-155-9

A catalogue record for this book is available from the National Library of New Zealand.

Book design: Megan van Staden
Printed in China by Everbest Printing Co. Ltd

BATEMAN BOOKS

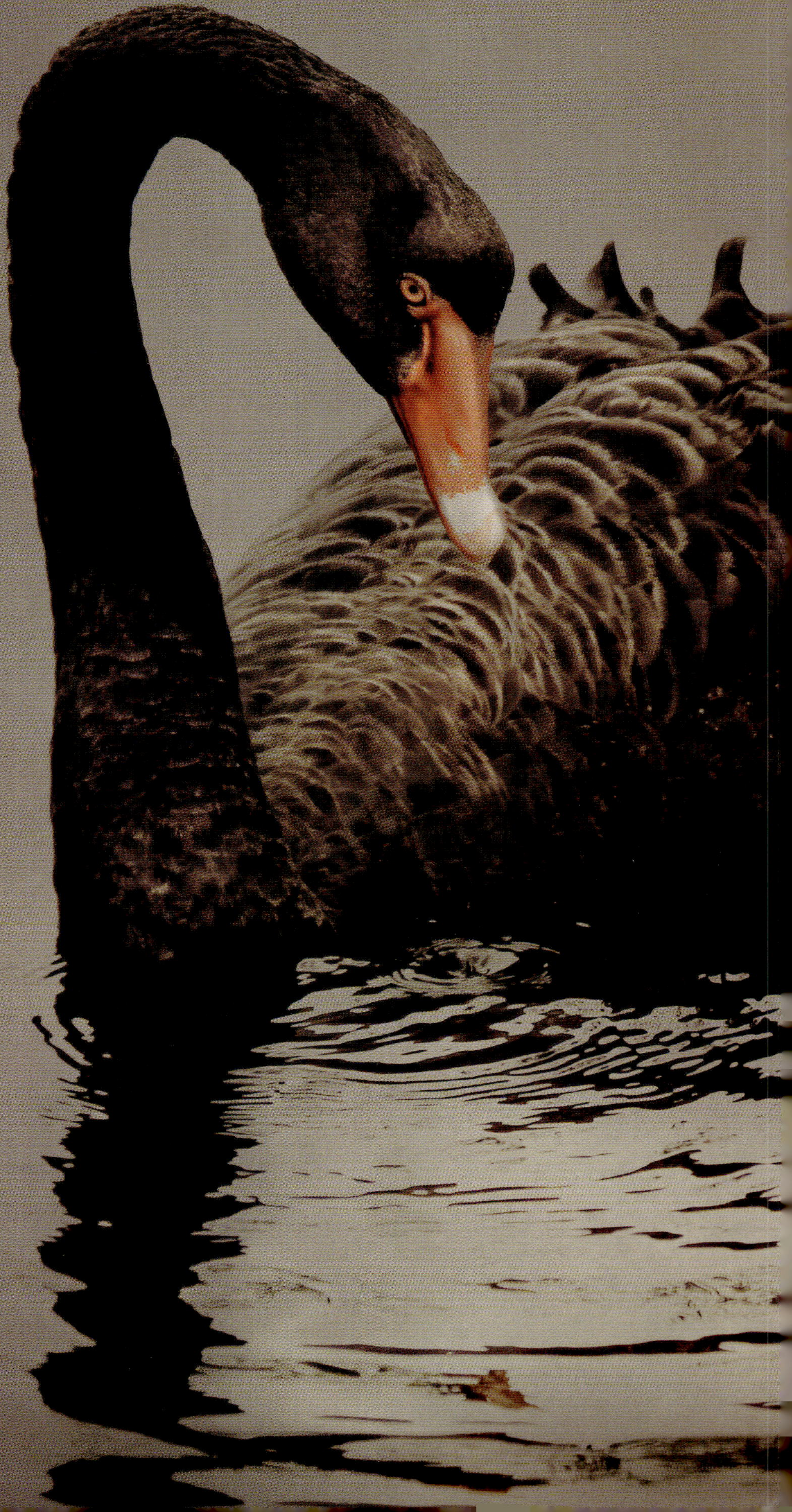